Progress in Applications of Boolean Functions

Synthesis Lectures on Digital Circuits and Systems

Editor
Mitchell A. Thornton, *Southern Methodist University*

Progress in Applications of Boolean Functions
Tsutomu Sasao and Jon T. Butler
2010

Embedded Systems Design with the Atmel AVR Microcontroller: Part II
Steven F. Barrett
2009

Embedded Systems Design with the Atmel AVR Microcontroller: Part I
Steven F. Barrett
2009

Embedded Systems Interfacing for Engineers using the Freescale HCS08 Microcontroller II: Digital and Analog Hardware Interfacing
Douglas H. Summerville
2009

Designing Asynchronous Circuits using NULL Convention Logic (NCL)
Scott C. Smith and Jia Di
2009

Embedded Systems Interfacing for Engineers using the Freescale HCS08 Microcontroller I: Assembly Language Programming
Douglas H. Summerville
2009

Developing Embedded Software using DaVinci & OMAP Technology
B.I. (Raj) Pawate
2009

Mismatch and Noise in Modern IC Processes
Andrew Marshall
2009

Pragmatic Circuits: Frequency Domain
William J. Eccles
2006

Pragmatic Circuits: Signals and Filters
William J. Eccles
2006

High-Speed Digital System Design
Justin Davis
2006

Introduction to Logic Synthesis using Verilog HDL
Robert B.Reese and Mitchell A.Thornton
2006

Microcontrollers Fundamentals for Engineers and Scientists
Steven F. Barrett and Daniel J. Pack
2006

© Springer Nature Switzerland AG 2022

Reprint of original edition © Morgan & Claypool 2010

Progress in Applications of Boolean Functions

Tsutomu Sasao and Jon T. Butler

ISBN: 978-3-031-79811-5 paperback
ISBN: 978-3-031-79812-2 ebook

DOI 10.1007/978-3-031-79812-2

A Publication in the Springer series
SYNTHESIS LECTURES ON DIGITAL CIRCUITS AND SYSTEMS

Lecture #26
Series Editor: Mitchell A. Thornton, *Southern Methodist University*
Series ISSN
Synthesis Lectures on Digital Circuits and Systems
Print 1932-3166   Electronic 1932-3174

Progress in Applications of Boolean Functions

Tsutomu Sasao
Fukuoka, Japan

Jon T. Butler
California, USA

SYNTHESIS LECTURES ON DIGITAL CIRCUITS AND SYSTEMS #26

ABSTRACT

This book brings together five topics on the application of Boolean functions. They are

1. Equivalence classes of Boolean functions: The number of n-variable functions is large, even for values as small as $n = 6$, and there has been much research on classifying functions. There are many classifications, each with its own distinct merit.

2. Boolean functions for cryptography: The process of encrypting/decrypting plaintext messages often depends on Boolean functions with specific properties. For example, highly nonlinear functions are valued because they are less susceptible to linear attacks.

3. Boolean differential calculus: An operation analogous to taking the derivative of a real-valued function offers important insight into the properties of Boolean functions. One can determine tests or susceptibility to hazards.

4. Reversible logic: Most logic functions are irreversible; it is impossible to reconstruct the input, given the output. However, Boolean functions that are reversible are necessary for quantum computing, and hold significant promise for low-power computing.

5. Data mining: The process of extracting subtle patterns from enormous amounts of data has benefited from the use of a graph-based representation of Boolean functions. This has use in surveillance, fraud detection, scientific discovery including bio-informatics, genetics, medicine, and education.

Written by experts, these chapters present a tutorial view of new and emerging technologies in Boolean functions.

KEYWORDS

logic functions, bent functions, cryptography, equivalence classes, nonlinear functions, Boolean difference, hazard detection, reversible logic, quantum computing, data mining, binary decision diagrams

Contents

Preface

ABOUT THIS BOOK

In the 1800s, Boolean logic was used to analyze human thought. It provided a mathematical basis for how conclusions could be made from observations. In the 1930's, Boolean logic was used to analyze and design telephone switching circuits and switch controllers for motors. However, after the invention of computers, it has mainly been used to represent logic circuits. Most textbooks on Boolean logic show optimization methods for two-level logic circuits. However, the scope of this book is different. The focus in on emerging areas of new applications. Written by experts in their respective fields, this book is suitable for graduate students working in computer science or electronic engineering.

Original papers were presented at the Reed-Muller Workshop, held in Naha, Okinawa, Japan, during May 23-24, 2009. After that, each chapter was modified to form a textbook, including the addition of examples and exercises. Significant effort has been devoted to making the presentation understandable by typical graduate students.

ORGANIZATION AND OVERVIEW

This book covers five topics.

- Recent advances in the understanding of various equivalence classes of Boolean functions. This is useful in applying the table-look up approach to logic synthesis.

- Cryptographically important Boolean functions. There exist desirable properties of Boolean functions that make decryption of encrypted messages difficult for unintended receivers.

- The Boolean difference operation. A conventional application has been test pattern generation for logic circuits. In addition, it applies to bi-decomposition and is useful for synthesis of multi-level logic circuits.

- Applications of reversible logic. Reversible logic is information lossless, meaning that the input can always be reconstructed from the output. Therewith, it builds the basis for applications e.g. in low-power design and quantum computation. Especially, the chapter introduces a design method for very large circuits.

- Data mining, where one seeks to extract hidden patterns from data. This is used in marketing, surveillance, fraud detection and scientific discovery for bio-informatics, genetics, medicine, and education. A graph-based representation of Boolean functions is used.

ACKNOWLEDGEMENTS

The authors and editors would like to thank the many reviewers who provided detailed comments on improving the chapters.

In addition to the editors, the chapters were reviewed by the following reviewers.
Chapter 1: Debatosh Debnath and Osnat Keren.

Chapter 2: D. Michael Miller and Claudio Moraga.

Chapter 3: Svetlana N. Yanushkevich and Mitchell A. Thornton.

Chapter 4: Igor L. Markov, Shigeru Yamashita, and Gerhard W. Dueck.

Chapter 5: Rolf Drechsler, Shinobu Nagayama, and Mohammed Zaki.

We are grateful to Mr. Munehiro Matsuura for editing the LATEX files to produce the camera-ready copy.

Tsutomu Sasao, Iizuka, Fukuoka, Japan
Jon T. Butler, Monterey, CA, USA
December 2009

CHAPTER 1

Equivalence Classes of Boolean Functions

Radomir S. Stanković, Stanislav Stanković, Helena Astola, and Jaakko T. Astola

CHAPTER SUMMARY

There are 2^{2^n} switching functions on n variables. Even for moderate values of n, this is a large number. To handle this complexity, n-variable switching functions are divided into classes under a set of selected classification rules. The rules are such that, once a representative function of a class is realized, the transformation to any other function in the class corresponds to modifications that are natural for the chosen technology. To be most useful, the classifications should have as few classes as possible, and the set of rules should be simple.

In this chapter, we review a few classifications including the classifications based on manipulation with Boolean expressions (NPN-classification), through LP-classification, SD-classification, and classification by Walsh spectra and Walsh decision diagrams.

1.1 INTRODUCTION

Classification of switching functions is one of the key problems in switching theory, because a classification, whose rules match the chosen representation of the function and the targeted technology, enables realization of all functions by simple modifications of a small set of circuit modules.

In switching theory, the term class of switching functions can denote functions that share some particular common properties. For example, linear, self-dual, monotone, threshold functions, etc. form particular and widely studied classes of switching functions [35], [37]. Consideration of such classes of functions is useful since, in general, functions with some peculiar properties are easier to analyze and realize than arbitrary switching functions. Threshold logic synthesis is a good example of this approach; see [1], [2], [37].

In another interpretation, a class of switching functions is a subset of the set of all 2^{2^n} switching functions of n variables whose elements can be derived from each other by applying some precisely defined classification rules. The following facts explain the continuing interest in classification of switching functions regarding various realization problems.

There are two major steps in logic synthesis:

1. For a given switching function, a multi-level network is determined, and then it is optimized with respect to various criteria as, for instance, the number of gates and testability. This is a technology-independent step [37].

2. The produced logic network is mapped to the targeted technology using a library of available physical cells or logic blocks. In such a library, usually, there are several cells capable of realizing the same function but with different area/speed characteristics. Optimization is performed again with respect to the area, speed, testability, dissipated power, etc.

Technology-independent network optimization, technology mapping, and cell library selection exploit the functional properties of the functions to be realized.

The goal of the classification is to find functions that are in some sense similar to each other. The notion of similarity is defined in different ways depending on the classification rules selected. For instance, one classification rule is just the complementation of the function values. In this case, two functions are equivalent if they are logic complements of each other. There are 2^{2^n-1} classes with 2 functions in each class. It follows that a function f and its logic complement $\overline{f}$ can be realized by adding an inverter to the output of the network that realizes f. The principle expressed in this example can be generalized. The functions that belong to the same class can be realized by similar networks. These networks are derived by simple modifications of the basic module that realizes the representative function of this class. The modifications correspond to the classification rules defining the classes.

Different classifications have been defined by selecting particular classification rules. In this chapter, we will briefly present the *NPN*-classification, *LP*-classification and *SD*-classification as examples of classification procedures based on different transformations of variables and function values. Then, we will consider classifications based on Walsh coefficients and Walsh decision diagrams.

1.2   CLASSIFICATION OF SWITCHING FUNCTIONS

Fig. 1.1 explains that a classification task consists of partitioning the set SF of all n-variable switching functions into classes C_i of functions. An exception is the SD-classification where the classes include functions with different number of variables. Each class is specified by a *representative function* c_i, the *representative* of the class C_i. Functions that belong to the same class can be obtained from each other by applying the operations performed in the classification. These operations are usually called *classification rules*. In particular, any function $f \in C_i$ can be transformed to the representative function c_i by the application of certain classification rules.

Each classification rule r is a 1-to-1 mapping of the set of all switching functions of n variables to itself, and a set of such rules $R = \{r_0, \ldots, r_k\}$ defines a classification. Two functions f_1 and f_2 belong to the same class if and only if there is a sequence of rules $t_1, \ldots, t_l \in R$ such

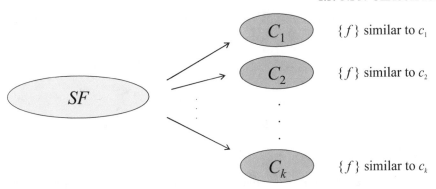

Figure 1.1: Classification of switching functions.

that $f_2 = t_l(t_{l-1}(\cdots(t_1(f_1))\cdots))$. As each function is in the same class with itself, it is natural to require that (1) the identity rule belongs to R. Similarly, it is natural to require that (2) if $r \in R$, then $r^{-1} \in R$. Recall that an equivalence relation ρ in a set X is such that it is

(a) Reflexive: $x\rho x$ (each element is in relation ρ with itself),

(b) Symmetric: If $x\rho y$, then $y\rho x$, and

(c) Transitive: If $x\rho y$ and $y\rho z$, then $x\rho z$.

Let $y \in X$. The set $\{x \in X | (x, y) \in \rho\}$ is called the equivalence class of X containing y. An equivalence relation partitions the set of functions into classes that are 1) disjoint, and 2) include all functions.

If we have a classification, as defined above by the set of rules $R = \{r_0, r_1, \ldots, r_k\}$, and we add a rule r and its inverse r^{-1}, where $r \notin R$, it can have the effect of merging of equivalence classes, yielding fewer classes.

1.3 NPN-CLASSIFICATION

Probably, the most widely used classification is the NPN-classification. In this case, simple modifications in the networks are required for a representative function to realize any function in the class.

In *NPN*-classification, the classification rules are

1. Negation (logical complement) of input variables (N) $\overline{x}_i \leftrightarrow x_i$,

2. Permutation of input variables (P) $x_i \leftrightarrow x_j$,

3. Negation of the output (N) $f \rightarrow \overline{f}$.

Depending on the allowed classification operations, the following subclassifications may be distinguished

1. N (rule 1),

2. P (rule 2),

3. NP (rules 1 and 2),

4. NPN (rules 1, 2 and 3);

NP and NPN-classification are most often used in practice. Classifications with a larger number of classification rules produce fewer classes and are considered to be *stronger*, since each representative function may represent a larger subset of functions.

We say that two functions f_1 and f_2 of the same number of variables n belong to the same class, or are equivalent, if they can be obtained from each other by the classification rules (N, P, NP, or NPN) allowed in the classification considered.

For instance, two functions are P-equivalent if they can be converted to each other by a permutation of variables. For a given number of variables, all functions that can be obtained from each other by a permutation of variables form a P-class. The NP-class is defined in a similar way, just by allowing negation of variables besides their permutation. If the truth-vector of a function f of n-variables is viewed as a 2^n bit binary number, then the corresponding integer is called the function number for f. In a P-class or an NP-class, we choose the function with the smallest function number to be the representative of the class.

Example 1.1   In Table 1.1, functions f_1 and f_2 are related by the negation of the variable x_3. The function f_3 is derived from f_1 by the permutation $x_1 \leftrightarrow x_3$. Thus, f_1 and f_3 are in the same P-class. Similarly, f_4 is derived from f_3 by the negation of x_1, and f_5 is derived from f_4 by negation of the variable x_3. Thus, functions f_1, f_2, f_3, f_4, and f_5 belong to the same NP-class. The function f_6 is negation of f_5. Therefore, all the mentioned functions belong to the same NPN-class. The other classes mentioned in the table will be discussed later in the corresponding sections.

The NP-classification has been considered by S.W. Golomb in 1959, [15], and later explored by many authors. Discussions of this topic can be found in [4], [10], [11], [20], [21], [22], [24], [35], [37], [41].

For $n = 3$, there are 14 NPN-representative functions, with ten of them dependent on all three variables. Table 1.3 shows these NPN-representatives.

Table 1.3 compares the number of equivalence classes under different classifications of functions. It shows the number of functions of n variables ($\# f$), functions that depend on all n variables ($\# f(n)$), and representative functions in C_P, C_{NP} and C_{NPN} classes. Interestingly, $\# f$ and $\# f(n)$ are nearly the same for large n.

Table 1.4 shows lower bounds on the number of classes.

Table 1.1: Examples of equivalent functions in different classifications.

Function f	F	Class
$f_1 = x_1 x_2 \vee x_3$	[01010111]	
$f_2 = x_1 x_2 \vee \overline{x}_3$	[10101011]	N-class with f_1
$f_3 = x_3 x_2 \vee x_1$	[00011111]	P-class with f_1
$f_4 = x_3 x_2 \vee \overline{x}_1$	[11110001]	NP-class with f_1, f_2, and f_3
$f_5 = \overline{x}_3 x_2 \vee \overline{x}_1$	[11110010]	NP-class with f_1, f_2, f_3
$f_6 = \overline{\overline{x}_3 x_2} \vee \overline{x}_1$	[00001101]	NPN-class with f_1, f_2, f_3, f_4
$f_7 = \overline{x}_1 \vee x_2 \vee x_3$	[11110111]	LP-class with f_1
$f_8 = \overline{x}_1 \overline{x}_3 \vee x_1 \overline{x}_2 \overline{x}_3 \vee x_1 x_2 x_3$	[10101001]	LP-class with f_1
$f_9 = x_1 \oplus (x_1 x_2 \vee x_3)$	[01011000]	W-class with f_1
$f_{10} = x_1 x_2 \vee x_2 x_3 \vee \overline{x}_2 \overline{x}_3$	[10011011]	WDD-class with f_1

Table 1.2: NPN-representatives for $n = 3$.

$NPN_1 = 1$	$NPN_8 = x_1 x_2 \vee \overline{x}_1 \overline{x}_2 x_3$
$NPN_2 = x_1$	$NPN_9 = x_1 x_2 \overline{x}_3 \vee x_1 \overline{x}_2 x_3 \vee \overline{x}_1 x_2 x_3$
$NPN_3 = x_1 \vee x_2$	$NPN_{10} = x_1 \overline{x}_2 \overline{x}_3 \vee \overline{x}_1 x_2 \overline{x}_3 \vee \overline{x}_1 \overline{x}_2 x_3 \vee x_1 x_2 x_3$
$NPN_4 = x_1 \oplus x_2$	$NPN_{11} = x_1 x_2 \vee x_1 x_3 \vee x_2 x_3$
$NPN_5 = x_1 x_2 x_3$	$NPN_{12} = x_1 \overline{x}_3 \vee x_2 x_3$
$NPN_6 = x_1 x_2 x_3 \vee \overline{x}_1 \overline{x}_2 \overline{x}_3$	$NPN_{13} = x_1 x_2 x_3 \vee x_1 \overline{x}_2 \overline{x}_3$
$NPN_7 = x_1 x_2 \vee x_1 x_3$	$NPN_{14} = x_1 x_2 \vee x_1 x_3 \vee \overline{x}_1 \overline{x}_2 \overline{x}_3$

All of the classifications mentioned in this section define an equivalence relation and, therefore, partition the set of all switching functions of a given number n of variables into disjoint classes. See [4], [20], [35], [37].

If the classification rules define an equivalence relation and the classification operations form a group, the exact number of equivalence classes can be determined by using combinatorial mathematics. For instance, some classification rules can be represented as matrix operations on the set of vectors corresponding to the switching functions. In some cases, those matrices form a group which acts on the set of functions represented as vectors. A group action assigns a permutation of the set to each element of the group. Then, by Cauchy-Frobenius Lemma (sometimes called Burnside's Lemma) [5], the number of equivalence classes is the average number of fixed points of the set under the group action.

Example 1.2   Consider the switching functions with $n = 2$ variables. The number of equivalence classes under P-classification is easily calculated using Cauchy-Frobenius Lemma.

We have a group of permutation matrices acting on the set of truth vector representations of switching functions. The group action is just regular matrix multiplication. The set of truth vectors

Table 1.3: Number of functions of n variables ($\#f$), functions dependent on all n variables ($\#f(n)$), and representative functions in C_P, C_{NP} and C_{NPN} classes.

	1	2	3	4	5	6
$\#f$	4	16	256	65,536	4.3×10^9	1.8×10^{19}
$\#f(n)$	2	10	218	64,594	4.3×10^9	1.8×10^{19}
$|C_P|$	4	12	80	3,984	3.7×10^7	2.6×10^{16}
$|C_{NP}|$	3	6	22	402	1,228,158	4.0×10^{14}
$|C_{NPN}|$	2	4	14	222	616,126	2.0×10^{14}

Table 1.4: Lower bounds on the number of classes.

Classification	P	NP	NPN
# of classes	$\dfrac{2^{2^n}}{n!}$	$\dfrac{2^{2^n}}{2^n n!}$	$\dfrac{2^{2^n}}{2^{n+1} n!}$

consists of 16 vectors of the form $\mathbf{F} = [f_0, f_1, f_2, f_3]^T$. The group of permutation matrices consists of two 4×4 matrices, the identity matrix $\mathbf{P_1}$ and the matrix corresponding to the interchange of the two variables of the switching functions $\mathbf{P_2}$:

$$\mathbf{P_1} = \begin{bmatrix} 1 & 0 & 0 & 0 \\ 0 & 1 & 0 & 0 \\ 0 & 0 & 1 & 0 \\ 0 & 0 & 0 & 1 \end{bmatrix}, \ \mathbf{P_2} = \begin{bmatrix} 1 & 0 & 0 & 0 \\ 0 & 0 & 1 & 0 \\ 0 & 1 & 0 & 0 \\ 0 & 0 & 0 & 1 \end{bmatrix}.$$

It is easily seen that permuting the two variables corresponds to interchanging f_1 and f_2 in the truth vector representations.

Now, multiplying by the identity matrix $\mathbf{P_1}$ leaves all the truth vectors fixed (16 vectors) and $\mathbf{P_2}$ leaves all the truth vectors having $f_1 = f_2$ fixed (8 vectors). The average of these fixed vectors is 12, which is the number of equivalence classes for $n = 2$ under P-classification.

The exact number of NP-classes has been computed up to $n = 8$ using Cauchy-Frobenius Lemma [3]. However, the groups quickly become very large, of size $2^n n!$ for n variables, and calculating the equivalence classes using this method is very time-consuming. In spite of that, it was possible to enumerate the NP-classes for $n = 6, 7$, and 8 as $4.0 \times 10^{14}, 5.3 \times 10^{32}$, and 1.1×10^{70}, respectively.

Another combinatorial approach to the problem is using the Pólya Enumeration Theorem, an extension to Cauchy-Frobenius Lemma, see for instance [47], to calculate the number of equivalence classes. This approach gives additional information on the structure of the groups, which can be further exploited for a deeper understanding of the equivalence relation. For further reading, see for instance [20], [21].

The relationships among *NPN*-classification and Reed-Muller expressions for switching functions were considered in [6]. The Reed-Muller expressions are a class of functional expressions to represent switching functions [4], [37]. They can be viewed as the decomposition of a given function $f(x_1, \ldots, x_n)$ into a linear combination of a set of basis functions defined by the elementary products of variables (the set of all possible products of switching variables x_i, $i = 1, \ldots, n$). The literals for the variables can be selected as either positive or negative, but not both at the same time in the same expression for f. This leads to *fixed-polarity Reed-Muller expressions (FPRMs)*, with the polarities for variables usually conveniently expressed by polarity vectors defined as n-bit vectors whose binary entries specify the selection of the positive and the negative literals for each of n variables. In this theory, the complexity of a Reed-Muller expression is often estimated in terms of two *weight vectors*.

The weight w_p of a Reed-Muller expression is defined as weight vector the number of product terms in the given expression for a specified polarity vector. Similarly, the weight w_l is the number of literals required in a Reed-Muller expression of a given polarity. The weight vector W_p (W_l) is a vector of all 2^n weights w_p (w_l) arranged in ascending order of their magnitudes.

In [6] and [7], it is shown that vectors W_p and W_l are invariant to *NPN*-classification operations.

In [46], see also [9], is proposed an analysis of different types of functions expressing certain properties that appear to be important and useful in circuit synthesis using macro cells [12]. Splitting switching functions into subset of functions sharing certain properties is motivated by various problems in technology mapping. For instance, in both mask and field programmable technologies, the minimization of the number of inverters is important [25]. Library cells with linear variables or cells that are capable of realizing self-complementary functions are useful in reducing the number of inverters by allowing them to shift between the inputs and outputs of the cell. To perform such an operation, the required property should be identified in advance. These sets of functions are efficiently described by representatives expressed in terms of *fixed-polarity Reed-Muller expressions* [46].

1.4   LP-CLASSIFICATION

NPN-classification is intended for AND-OR synthesis, which means the representation of functions by AND-OR expressions, usually called SOPs. From 1990, there has been an increasing interest in AND-EXOR synthesis, mainly after the publication of [38], where it was shown that AND-EXOR expressions, usually called ESOPs, require on the average fewer products. Another factor was the technology advance that provided EXOR circuits with the same propagation delay and at about the same price as classical OR circuits with the same number of inputs.

t	AND-OR	AND-EXOR
Table 1.5: Number of functions realizable by t products [28].		
0	1	1
1	81	81
2	1,804	2,268
3	13,472	21,744
4	28,904	37,530
5	17,032	3,888
6	3,704	24
7	512	0
8	26	0
avg.	4.13	3.66

Example 1.3   [28] Table 1.5 shows the number of functions of four variables that can be realized with t products by AND-OR and AND-EXOR expressions. Notice that AND-OR expressions require on the average 13% more products than AND-EXOR expressions.

These facts have been behind the introduction of *LP-classification* where the classification rules have been adapted for AND-EXOR synthesis [36], both for binary and quaternary cases. The theory has been elaborated further in [28], [29], [30], [31]. The key property of LP-classification rules is that they do not change the number of product terms in ESOPs. In this classification, unlike NPN-classification, transformations over constants, in addition to variables, are also allowed.

The binary LP-classes are defined as equivalence classes under the following transformations that change a function f to (possibly) another function g

1. For $i \in \{1, \ldots, n\}$ and $f(x_1, \ldots, x_n)$ written in the Shannon expansion with respect to the variable x_i

$$f = \overline{x}_i f_0 \oplus x_i f_1.$$

the function g is in the form

$$g = \begin{bmatrix} g_0 \\ g_1 \end{bmatrix} = \begin{bmatrix} a & b \\ c & d \end{bmatrix} \begin{bmatrix} f_0 \\ f_1 \end{bmatrix},$$

i.e.,

$$g = \overline{x}_i g_0 \oplus x_i g_1 = \overline{x}_i (af_0 \oplus bf_1) \oplus x_i (cf_0 \oplus df_1),$$

where $\begin{bmatrix} a & b \\ c & d \end{bmatrix}$ is a nonsingular matrix over $GF(2)$.

2. g is obtained from f by permuting variables.

Functions f and g are called LP-equivalent if g is obtained from f by a sequence of (operations) transformations 1. and 2. above [31].

It is clear that LP-equivalence is an equivalence relation. The exact number of LP-classes has been computed up to $n = 6$ using Cauchy-Frobenius Lemma [3].

Example 1.4   Consider the transformation by the matrix $\begin{bmatrix} 1 & 0 \\ 1 & 1 \end{bmatrix}$. Then,

$$g = \begin{bmatrix} g_0 \\ g_1 \end{bmatrix} = \begin{bmatrix} 1 & 0 \\ 1 & 1 \end{bmatrix} \begin{bmatrix} f_0 \\ f_1 \end{bmatrix} = \begin{bmatrix} f_0 \\ f_0 \oplus f_1 \end{bmatrix},$$

and equivalently

$$g = \bar{x}_i g_0 \oplus x_i g_i = \bar{x}_i f_0 \oplus x_i (f_0 \oplus f_1) = (\bar{x}_i \oplus x_i) f_0 \oplus x_i f_1 = 1 \cdot f_0 \oplus x_i f_1.$$

Thus, g is obtained by the substitution $\bar{x}_i \leftrightarrow 1$ in the expression of f.

Similarly, the matrix $\begin{bmatrix} 1 & 1 \\ 0 & 1 \end{bmatrix}$ corresponds to the substitution $x_i \leftrightarrow 1$.

Usually, the six transformations corresponding to the six different nonsingular matrices over $GF(2)$ are expressed in this substitution notation as shown in Table 1.6.

Table 1.6: *LP*-classification rules.		
1. $LP_0(f) = \bar{x}_i f_0 \oplus x_i f_1$	Identity mapping	$\begin{bmatrix} 1 & 0 \\ 0 & 1 \end{bmatrix}$
2. $LP_1(f) = \bar{x}_i f_0 \oplus 1 \cdot f_1$	$x_i \leftrightarrow 1$	$\begin{bmatrix} 1 & 1 \\ 0 & 1 \end{bmatrix}$
3. $LP_2(f) = 1 \cdot f_0 \oplus x_i f_1$	$\bar{x}_i \leftrightarrow 1$	$\begin{bmatrix} 1 & 0 \\ 1 & 1 \end{bmatrix}$
4. $LP_3(f) = x_i f_0 \oplus \bar{x}_i f_1$	$x_i \leftrightarrow \bar{x}_i$	$\begin{bmatrix} 0 & 1 \\ 1 & 0 \end{bmatrix}$
5. $LP_4(f) = x_i f_0 \oplus 1 \cdot f_0$	$x_i \to 1$, and $\bar{x}_i \to x_i$	$\begin{bmatrix} 0 & 1 \\ 1 & 1 \end{bmatrix}$
6. $LP_5(f) = 1 \cdot f_0 \oplus \bar{x}_i f_1$	$\bar{x}_i \to 1$, and $x_i \to \bar{x}_i$	$\begin{bmatrix} 1 & 1 \\ 1 & 0 \end{bmatrix}$

Example 1.5   The function f_1 in Table 1.1 can be written as an AND-EXOR expression $f_1 = x_3 \oplus x_1 x_2 \bar{x}_3$. The transformations $x_3 \to 1$ and $x_2 \to \bar{x}_2$ convert f_1 into the function $f_7 = 1 \oplus x_1 \bar{x}_2 \bar{x}_3 = \bar{x}_1 \vee x_2 \vee x_3$. These transformations are defined by transform matrices in the rows 2 and

4 in Table 1.6. Therefore, the truth-vector of f_7 can be derived from the truth-vector of f_1 by a transformation matrix that is the Kronecker product of the matrices in rows 1, 4, and 2, corresponding to the transforms performed. Thus, the truth-vector for f_7 is $\mathbf{F}_7 = [11110111]^T$.

The transformation $\overline{x}_3 \rightarrow 1$ and $x_3 \rightarrow \overline{x}_3$, defined by the transform matrix in the row 6 in Table 1.6, converts f_1 into the function $f_8 = \overline{x}_3 \oplus x_1 x_2 = \overline{x}_1 \overline{x}_3 \vee x_1 \overline{x}_2 \overline{x}_3 \vee x_1 x_2 x_3$. The truth-vector $\mathbf{F}_8 = [10101001]^T$ is determined from the truth-vector for f_1 by the matrix, which is the Kronecker product of two identity matrices of order 2 and the matrix in the row 2 of Table 1.6.

Therefore, functions f_1, f_7, and f_8 belong to the same LP-class, although not to the NPN-class, due to the replacement of a literal by the constant 1 which is not an operation allowed in NPN-classification rules.

For functions of $n = 3$ variables, there are 6 LP-representative functions, which are shown in Table 1.7.

Table 1.7: LP-representatives for $n = 3$.	
$LP_1 = 0$	$LP_4 = \overline{x} \cdot \overline{y} \cdot \overline{z}$
$LP_2 = \overline{x} \cdot y \oplus \overline{x} \cdot z$	$LP_5 = \overline{x} \oplus \overline{y} \cdot z \oplus \overline{x} \cdot y \cdot z$
$LP_3 = x \cdot \overline{y} \cdot \overline{z} \oplus \overline{x} \cdot y \cdot z$	$LP_6 = \overline{x} \oplus y \cdot \overline{z} \oplus x \cdot \overline{y} \cdot z$

As in the case of the NPN-classification, the LP-representative functions are selected by the smallest function numbers and represented by the minimum ESOPs (MESOPs) [30].

Table 1.8 shows the number of LP-representative functions for up to 6 variables, which can be compared to the data in Table 1.3. Since, in any class, there are at most $n!6^n$ functions, it follows that the number of LP-classes is at least $2^{2^n}/n!6^n$. It should be noticed that LP-classification is stronger than NPN-classification since it considerably reduces the number of classes.

To decide the equivalence class of a given function by the definition of LP-equivalence, requires a computation complexity that is proportional to $n!6^n$. However, this complexity can be reduced to $n3^n$ ($n \leq 5$) if the LP-characteristic vectors defined in [28] are used to specify the LP-equivalence classes.

Table 1.8: Number of LP-classes and its medium sizes.						
n	1	2	3	4	5	6
$|C_{LP}|$	2	3	6	30	6,936	5.4996×10^{11}
#f	4	16	256	65,536	4.3×10^9	1.8×10^{19}
#$f/|C_{LP}|$	2	5.3	42.6	2,184.5	6.2×10^5	3.3×10^7

1.5   SD-CLASSIFICATION

Another important classification of switching functions in the Boolean domain, is defined in terms of self-dual functions and is, therefore, called the self-dual (SD) classification of switching functions. The present interest in this classification is due to a close relationship with threshold logic, which is important in the realization of neural networks [1].

For a function $f(x)$, $x = (x_1, x_2, \ldots, x_n)$, the dual $f^d(x)$ is $f^d(x) = \overline{f(\overline{x})}$. For example, if $f(x_1, x_2, x_3, x_4) = x_1\overline{x}_2 \vee \overline{x}_3 x_4$, then $f^d(x_1, x_2, x_3, x_4) = \overline{(\overline{x}_1 x_2 \vee x_3 \overline{x}_4)}$. If $f^d(x) = f(x)$, then the function $f(x)$ is said to be self-dual. For example, the majority function is a self-dual function, as can be seen for $n = 3$,

$$
\begin{aligned}
f^d(x) &= \overline{\overline{x}_1\overline{x}_2 \vee \overline{x}_2\overline{x}_3 \vee \overline{x}_1\overline{x}_3} \\
&= (x_1 \vee x_2)(x_2 \vee x_3)(x_1 \vee x_3) \\
&= x_1 x_2 \vee x_2 x_3 \vee x_1 x_3 = f(x).
\end{aligned}
$$

Given a self-dual function $f^d(x)$ of n variables, we can create a self-dualized function [16], [17] on $n + 1$ variables as

$$
f^h(x) = x_{n+1} f(x) \vee \overline{x}_{n+1} f^d(x), \tag{1.1}
$$

where $x_{n+1} \notin x$.

Similarly, a switching function of n variables associated with a self-dual switching function of $(n + 1)$ variables by $f = f^{sd}(x_i = 1)$, i.e., by specifying $x_i = 1$ in (1.1), is called the *anti-self-dualized function* of f^{sd}.

A self-dualized function is a self-dual function, as

$$
\begin{aligned}
f^d(f^h(x)) &= \overline{(\overline{x}_{n+1} f(\overline{x}) \vee \overline{x}_{n+1} f^d(\overline{x}))} \\
&= \overline{(\overline{x}_{n+1} f(\overline{x}))} \, \overline{(x_{n+1} f^d(\overline{x}))} \\
&= (x_{n+1} \vee \overline{f(\overline{x})})(\overline{x}_{n+1} \vee \overline{f^d(\overline{x})}) \\
&= (x_{n+1} \vee f^d(x))(\overline{x}_{n+1} \vee f(x)) \\
&= x_{n+1} f(x) + \overline{x}_{n+1} f^d(x) + f(x) f^d(x) \\
&= x_{n+1} f(x) \vee \overline{x}_{n+1} f^d(x) = f^h(x).
\end{aligned}
$$

The self-dualized $f^h(x)$ of any self-dual function $f(x)$ is the function $f(x)$, since $x_{n+1} f(x) \vee \overline{x}_{n+1} f^d(x) = f(x)(x_{n+1} \vee \overline{x}_{n+1}) = f(x)$.

A function $f(x)$ and its dual $f^d(x)$ are obviously in the same *NPN*-class, since $f^d(x)$ is defined by negation of the output and the variables. The self-dualized function of $f(x)$, however, does not belong to the same *NPN*-class as its dual, unless it is self-dual.

Any self-dualized function of $(n + 1)$ variables can be decomposed into two non-self dual functions of n variables with respect to any single variable. A self-dual function of $(n + 1)$ variables and all the possible decompositions into functions of n variables are in the same SD-class. Further, *NPN*-classification operations can be applied to all the functions within the same SD-class without moving outside the given SD-class.

By using these concepts, the SD-classes are defined in [17] as follows:

- A class of switching functions is called an SD-class if the class consists of switching functions (non-self-dual functions of n variables, and self-dual functions of $(n + 1)$ variables) that coincide one to another by self-dualization, anti-self-dualization, negation of functions and negation and permutation of variables.

- These operations are called the SD-classification operations and can be applied in an arbitrary order and as many times as required to convert a function into another targeted function in the same class. Thus, the SD-classification is the NPN-classification enriched with two additional classification rules, the self-dualization rule and the anti-self-dualization rule.

- Functions f and f^d belong to the same NPN-class by appropriate input and output negation operations. A self-dualized function f^{sd} of $(n + 1)$ variables may cover several dissimilar NPN-representatives of n variables under the Shannon decomposition with respect to each variable x_i. Therefore, a self-dualized function f^{sd} of $(n + 1)$ variables may subsume several NPN-representatives of n variables resulting in this way in a stronger classification of switching function. For instance, when $n \leq 4$, there are 83 SD-classes compared to 222 NPN-classes.

Table 1.9 shows the SD-representatives for $n = 3$.

Table 1.9: SD-representatives for $n = 3$.
$SD_1 = x_1$
$SD_2 = x_1 x_2 \vee x_2 x_3 \vee x_1 x_3$
$SD_3 = x_1 \oplus x_2 \oplus x_3$
$SD_4 = x_1 x_2 x_3 \vee (x_1 \vee x_2 \vee x_3) x_4$
$SD_5 = (x_1 x_2 x_3 \vee \overline{x}_1 \overline{x}_2 \overline{x}_3) x_4 \vee (x_1 \vee x_2 \vee x_3)(\overline{x}_1 \vee \overline{x}_2 \vee \overline{x}_3) \overline{x}_4$
$SD_6 = x_1 (x_2 x_3 \vee \overline{x}_2 \overline{x}_3) x_4 \vee (x_1 \vee x_2 \overline{x}_3 \vee \overline{x}_2 x_3) \overline{x}_4$
$SD_7 = (\overline{x}_1 x_2 x_3 \vee x_1 \overline{x}_2 x_3 \vee x_1 x_2 \overline{x}_3) \overline{x}_4 \vee (x_1 x_2 \vee x_2 x_3 \vee x_1 x_3 \vee \overline{x}_1 \overline{x}_2 \overline{x}_3) \overline{x}_4$

The SD-classification was introduced [17] as a classification that plays the same role in *threshold logic synthesis* as that of the NPN-classification does in *AND–OR* synthesis.

$f(x)$ is a *threshold function* if

$$f(x) = \begin{cases} 1, & \text{if } \sum_{i=1}^{n} w_i x_i \geq T, \\ 0, & \text{if } \sum_{i=1}^{n} > T, \end{cases}$$

where $w_i, i = 1, \ldots, n$ are real-valued *weights* and T is the real-valued *threshold*.

Elementary logic gates such as *AND, OR,* and *NOT* are special cases of threshold gates. For the *AND* gate $w_1 = w_2 = \cdots = w_n = 1$ and $T = n$. For the *OR* gate, $w_1 = w_2 = \cdots = w_n = T = 1$. The single input threshold gate with $w_1 = -1$ and $T = 0$ realizes the *NOT* gate. Since these elementary logic gates are a complete set, meaning that any switching function can be implemented by a network of *AND, OR,* and *NOT* gates, it follows that any switching function can be also implemented by a network of threshold gates. In threshold logic, each threshold gate, in general, realizes a more complex function than elementary *AND, OR, NOT* and similar gates in conventional logic. Consequently, the number of threshold gates in a threshold network implementing a function is never larger and often smaller than the number of gates in a network using conventional gates. A similar remark can be given regarding the number of levels, which determines the speed of the network, and the number of interconnections, which determines the density of the network. It does not mean that the threshold logic would always produce faster and more economical realizations. A proper comparison of various performances is a complex task and depends on the particular technologies used to implement gates. Recent interest in threshold logic synthesis, and indirectly in SD-classification, is due to the advent of electronic gates as for instance resonant tunneling diodes (RTDs) and quantum cellular automata (QCAs) that are capable of realizing threshold logic, see for instance [18], [19], [51].

1.6   CLASSIFICATION BY WALSH COEFFICIENTS

In the 70's, there was a wide interest in *discrete Walsh functions*. There are applications in many areas of computing and information technologies based on discrete Walsh functions. In this section, these functions will be used for classification of switching functions.

1.6.1   WALSH FUNCTIONS

Discrete Walsh functions $wal(w, x)$, $w, x \in \{0, 1, \ldots, 2^n - 1\}$ can be viewed as columns of the $(2^n \times 2^n)$ *Walsh matrix* recursively defined as

$$\mathbf{W}(n) = \left[\begin{array}{cc} W(n - 1) & W(n - 1) \\ W(n - 1) & -W(n - 1) \end{array} \right], \quad \mathbf{W}(1) = \left[\begin{array}{cc} 1 & 1 \\ 1 & -1 \end{array} \right],$$

Since this is a non-singular matrix, the set of 2^n discrete Walsh functions forms a *basis* in the space of functions defined in 2^n points, i.e., functions represented by function vectors of length 2^n. Being a basis means that each of the functions in the considered space can be represented as a linear combination of the discrete Walsh functions. The *Walsh spectrum* of an n-variable function is obtained by multiplying the function vector of length 2^n by the $(2^n \times 2^n)$ Walsh matrix $\mathbf{W}(n)$. Usually, so-called $(1, -1)$ encoding is used for the function vector, meaning that each 0 is replaced by -1 and each 0 by 1. As the transform is performed in real arithmetic, this is more natural.

The Walsh transform is a self-inverse transform up to the constant 2^{-n} that is used as the normalization factor when defining the Walsh transform and its inverse.

For instance, it can be shown that, with this encoding, the Walsh coefficients are even integers in the range 2^{-n} to 2^n. Further, not every combination of such integers can appear in the Walsh spectrum of a switching function due to their binary values, and the sum of Walsh coefficients of a switching functions has to always be equal to 2^n.

Normalizing the coefficients in a Walsh transform are important for the correct representation of a concrete function. However, it is irrelevant for the classification of it. If we discard the normalizing coefficient, the Walsh transform of a function $f(\mathbf{x}), \mathbf{x} = (x_1, \ldots, x_n), x_i \in \{0, 1\}$, can be written as

$$\mathbf{W}_f(\mathbf{u}) = \sum_{\mathbf{x} \in B_n} (-1)^{\mathbf{u}^T \mathbf{x}} f(\mathbf{x}), \tag{1.2}$$

where $\mathbf{u} = (u_1, \ldots, u_n), u_i \in \{0, 1\}$, and

$$(-1)^{\mathbf{u}^T \mathbf{x}} = \begin{cases} -1, & \text{if } \mathbf{u}^T \mathbf{x} = 1, \\ 1, & \text{if } \mathbf{u}^T \mathbf{x} = 0. \end{cases}$$

Example 1.6   Consider the function $f_1 = x_1 x_2 \vee x_3$ in Table 1.1. The truth-vector of f_1 is $\mathbf{F}_1 = [0, 1, 0, 1, 0, 1, 1, 1]^T$ and the encoding $\{0, 1\} \rightarrow \{-1, 1\}$ gives $\mathbf{F}_1 = [1, -1, 1, -1, 1, -1, -1, -1]$. The Walsh spectrum for f_1 is calculated as $\mathbf{W}_{f_1} = [-2, 6, 2, 2, 2, 2, -2, -2]^T$.

The Walsh functions of the first order, i.e., of the index 2^i, $i = 1, \ldots, n$, are identical to the Rademacher functions $r_i(x)$, $i \in \{0, 1, \ldots, n\}$, $x \in \{0, 1, \ldots, 2^n - 1\}$. The Walsh functions of higher others can be generated as the products of Rademacher functions. By using Kronecker product, the set of all 2^n Walsh functions is generated as

$$\bigotimes_{i=1}^{n} [\, 1 \quad r_i(x) \,].$$

For example, when $n = 2$ the functions are

$$[1 \quad r_1(x)] \otimes [1 \quad r_2(x)] = [1 \quad r_2(x) \quad r_1(x) \quad r_1(x) r_2(x)].$$

To emphasize this correspondence between the Rademacher and Walsh functions, the Walsh coefficients are often denoted by $r_{i,j,\ldots,k}$ where i, j, k are indices of the corresponding Rademacher functions. We say that the Walsh functions generated in this way are in the *Hadamard ordering*, and the Walsh coefficients are also in the same ordering. For example, if $n = 4$, the Walsh coefficients in the Hadamard ordering are

$$\mathbf{W}_f = [r_0, r_4, r_3, r_{34}, r_2, r_{24}, r_{23}, r_{234}, r_1, r_{14}, r_{13}, r_{134}, r_{12}, r_{124}, r_{123}, r_{1234}]^T. \tag{1.3}$$

For classification purposes, and in some other applications [22], the coefficients are usually ordered in the increasing number of subscripts and the increasing order of their values. For example, if $n = 4$, the Walsh coefficients are ordered as

$$\mathbf{W}_f = [r_0, r_1, r_2, r_3, r_4, r_{12}, r_{13}, r_{14}, r_{23}, r_{24}, r_{34}, r_{123}, r_{124}, r_{134}, r_{234}, r_{1234}]^T. \tag{1.4}$$

The convenience of this notation comes from the fact that certain properties of switching functions can be characterized uniquely by the first $n + 1$ coefficients. For instance, threshold functions are uniquely characterized by $(n + 1)$ coefficients called the Chow parameters or modified Chow parameters [8]. These parameters are defined in slightly different ways by different authors; however, in all these definitions, the zeroth and first order Walsh coefficients are identical, up to a scaling factor [22]. In connection with this, classification in terms of all 2^n Walsh spectral coefficients has been considered [13], [14], [22].

Walsh functions of the first order correspond to switching variables in $\{0, 1\} \rightarrow \{1, -1\}$ encoding, and the Walsh functions of higher orders correspond to linear combinations of switching variables, i.e., linear switching functions. This feature has been used efficiently in several tasks in switching theory. Since Walsh coefficients of higher orders are defined in terms of products of first order Walsh functions, they are mutually dependent in the sense that changing a coefficient of the first order will cause the corresponding changes in certain coefficients of higher order as will be discussed below. There are certain operations in the Boolean domain that result in permutation and change of signs of Walsh coefficients which, however, do not change the absolute values. These operations are called *translation invariant spectral operations* [22] and will be used as the classification operations below.

1.6.2   CLASSIFICATION BY USING WALSH FUNCTIONS

In classification by Walsh coefficients (W-classification), the following classification rules, which correspond to the translation invariant spectral operations, are allowed:

1. Interchange of variables x_i and x_k, $i \neq k$,

2. Replacement of a variable by its complement $\overline{x}_i \rightarrow x_i$,

3. Replacement of a function f by its complement $\overline{f}$,

4. Replacement of a variable x_i by $x_i \oplus x_k$, which is called *spectral translation*,

5. Replacement of a function $f(x_1, \ldots, x_n)$ by $x_i \oplus f(x_1, \ldots, x_n)$, which is called the *disjoint spectral translation*.

The operation $\overline{x}_i \rightarrow x_i$ can be written as $x_i \oplus 1 \rightarrow x_i$, or over all variables as $\mathbf{x} \oplus \mathbf{v} \rightarrow \mathbf{x}$, where $\mathbf{x} = (x_1, \ldots, x_n)$, $\mathbf{v} = (v_1, \ldots, v_n) \in \{0, 1\}^n$, and $\oplus$ is applied componentwise. Similarly, the operation $x_i \oplus x_k \rightarrow x_i$ can be expressed as $\mathbf{A}\mathbf{x} \rightarrow \mathbf{x}$, where $\mathbf{A}$ is an $(n \times n)$ matrix over $GF(2)$ such that $a_{st} = 1$ if $s = t$, $a_{ik} = 1$ and $a_{st} = 0$ otherwise.

In the spectral domain, the above classification operations can be formulated using matrix operations as follows. This formulation is useful for defining the corresponding algorithms and their programming implementations.

Then, the spectral invariance properties of rules 2-5 can be interpreted as

2. $g(\mathbf{x}) = f(\mathbf{x} \oplus \mathbf{v}) \Rightarrow \mathbf{W}_g(\mathbf{u}) = (-1)^{\mathbf{u}^T \mathbf{v}} \mathbf{W}_f(\mathbf{u})$,

3. $g(\mathbf{x}) = -f(\mathbf{x}) \Rightarrow \mathbf{W}_g(\mathbf{u}) = -\mathbf{W}_f(\mathbf{u})$,

4. $g(\mathbf{x}) = f(\mathbf{Ax}) \Rightarrow \mathbf{W}_g(\mathbf{u}) = \mathbf{W}_f(\mathbf{A}^T\mathbf{u})$,

5. $g(\mathbf{x}) = f(\mathbf{x})(-1)^{\mathbf{v}^T\mathbf{x}} \Rightarrow \mathbf{W}_g(\mathbf{u}) = \mathbf{W}_f(\mathbf{u} \oplus \mathbf{v})$.

In other words, spectral translation states that, given a function and its spectrum, a linear (mod (2)) transformation of arguments corresponds to a particular permutation of spectral coefficients. The disjoint spectral translation states that, given a function f and its Walsh spectrum W_f, the spectrum of a new function f' obtained by modulo 2 addition of the original function and some of its arguments, corresponds to a permutation and/or change of signs in the Walsh spectrum W_f. These permutations in the spectrum can also be characterized by certain manipulations of the indices of Walsh coefficients [22], [24], [34].

The above operations have been used to define classes of *disjointly translationally equivalent functions*, W-classes, as functions which can be derived from each other by rules 1-5.

The following Example 1.7 presents W-equivalent functions.

Example 1.7   The following functions all belong to the same class in classification with respect to Walsh coefficients (W-class) with function f_1 in Example 1.6. The Walsh spectrum of f_1 with the scaling factor 2^{-3} omitted is $\mathbf{W}_{f_1} = [-2, 6, 2, 2, 2, 2, -2, -2]^T$.

The function f_2 in Table 1.1 is obtained from f_1 by the negation of x_3. The Walsh spectrum for f_2 can be calculated by using the equation in spectral invariance property 2, in which case $\mathbf{v} = [0, 0, 1]$ and $\mathbf{W}_{f_2} = [-2, -6, 2, -2, 2, -2, -2, 2]$.

For the function $\overline{f}_1 = \overline{x}_1\overline{x}_3 \vee \overline{x}_2\overline{x}_3$, the Walsh spectrum is, by spectral invariance property 3, $\mathbf{W}_{\overline{f}_1} = [2, -6, -2, -2, -2, -2, 2, 2]^T$.

If the disjoint spectral translation $x_1 \oplus f_1 \to f_1$ is applied, the function $f_9 = x_1 \oplus (x_1x_2 \vee x_3)$ is obtained. The Walsh spectrum for this function can be calculated by spectral invariance property 5, where now $\mathbf{v} = [1, 0, 0]$, and the Walsh spectrum is then $\mathbf{W}_{f_9} = [2, 2, -2, -2, -2, 6, 2, 2]$.

In terms of the manipulation of indices, that is useful for understanding the related classification procedures, and is used to define the representative of W-classes, the classification operations can be formulated as follows [22]. It is assumed that coefficients are denoted and ordered as in (1.4), i.e., first the Walsh coefficients of the first order and then coefficients of higher orders as determined by the products of the Rademacher functions.

1. Negation of a variable x_i, $i \neq 0$, results in the change of the sign of Walsh coefficients containing the index i, as for instance r_i, r_{ik}, r_{ijk}, etc.

2. Permutation of variables x_i and x_j, $i \neq j \neq 0$, results in the interchange of all pairs of coefficients which contain i and j in their subscripts. For instance, $r_i \leftrightarrow r_j, r_{ik} \leftrightarrow r_{jk}$, etc.

3. Negation of the function f results in the change of the sign of all the coefficients including r_0.

4. Spectral translation x_i by $x_i \oplus x_j, i \neq j \neq 0,$ results in an interchange of all pairs of coefficients which contain i in their indices. This interchange can be defined as *in all coefficients that contain i in the index delete j if it also appears and append it if it does not*. For example, $r_i \leftrightarrow r_{ij}, r_{ik} \leftrightarrow r_{ijk},$ etc.

5. Disjoint spectral translation $f \rightarrow f \oplus x_i,$ results in an interchange of the coefficients which can be described as *in all coefficients append i if does not appear and delete it if it is already present*. For example, $r_i \leftrightarrow r_0, r_{ij} \leftrightarrow r_j, r_{ijk} \leftrightarrow r_{jk},$ etc.

W-classes are specified by Walsh spectra in $(1, -1)$ encoding, in the ordering referring to the Rademacher functions as specified above, and with the first $n + 1$ coefficients, r_0 and first order coefficients $r_1, \ldots, r_n,$ with positive values. These spectra are *W-representatives*, and the first $n + 1$ coefficients form the *positive canonic vector* that is sufficient to characterize certain classes of switching functions as for instance linearly separable functions, and threshold functions mentioned above [22]. This definition of W-representatives is convenient, since it permits us to relate the positive canonic vectors with Chow parameters which are also used to describe the mentioned types of switching functions [8].

It has been shown that, for $n \leq 4,$ there are 8 classes of functions, seven of which are threshold functions. Threshold functions can be realized by a single threshold logic element [13] and, so any function in such a class can be realized by adding EXOR gates to this element.

For $n \leq 3,$ there are three W-representatives, which are shown in Table 1.10. They can be specified by the first four entries since these are unique for each classification entry.

Table 1.10: W-representatives for $n = 3$.	
$\mathbf{W}_{f_1} = [8, 0, 0, 0, 0, 0, 0, 0, 0]$	$f(x_1, x_2, x_3) = 0$
$\mathbf{W}_{f_2} = [6, 2, 2, 2, -2, -2, -2, 2]$	$f(x_1, x_2, x_3) = x_1 x_2 x_3$
$\mathbf{W}_{f_3} = [4, 4, 4, 0, -4, 0, 0, 0]$	$f(x_1, x_2, x_3) = x_1 x_2$

Table 1.11 shows the *W-representative functions* for $n \leq 4$ [22]. The rows 1,3, and 5 are for functions of $n < 4$ variables, while other rows correspond to functions that depend on all four variables. Table 1.12 shows these functions in the Boolean domain. The first two W-representatives are equal to the first two LP-representatives for $n \leq 4,$ while others are different. Table 1.13 shows the number of W-classes for $n = 1, 2, 3, 4, 5$ [22].

To show that a function belongs to a certain W-class, its Walsh spectrum should be converted into a W-representative. The following example illustrates the W-classification procedure.

Example 1.8   Consider a switching function f of $n = 4$ variables defined by the truth-vector

$$\mathbf{F} = [1, 1, 0, 1, 1, 1, 1, 1, 0, 1, 0, 1, 0, 0, 0, 1]^T.$$

	r_0	r_1	r_2	r_3	r_4	r_{12}	r_{13}	r_{14}	r_{23}	r_{24}	r_{34}	r_{123}	r_{124}	r_{134}	r_{234}	r_{1234}
1.	16	0	0	0	0	0	0	0	0	0	0	0	0	0	0	0
2.	14	2	2	2	2	−2	−2	−2	−2	−2	−2	2	2	2	2	−2
3.	12	4	4	4	0	−4	−4	0	−4	0	0	4	0	0	0	0
4.	10	6	6	2	2	−6	−2	−2	−2	−2	2	2	2	−2	−2	2
5.	8	8	8	0	0	−8	0	0	0	0	0	0	0	0	0	0
6.	8	8	4	4	4	−4	−4	−4	0	0	0	0	0	0	−4	4
7.	6	6	6	6	6	−2	−2	−2	−2	−2	−2	−2	−2	−2	−2	6
8.	4	4	4	4	4	4	4	−4	−4	4	−4	−4	−4	4	4	−4

Table 1.11: W-representatives for $n \leq 4$.

Table 1.12: W-representatives for $n \leq 4$ in the Boolean domain.

$w_1 = 0$	$w_5 = x_1 x_2$
$w_2 = x_1 x_2 x_3 x_4$	$w_6 = x_1 x_2 x_3 \lor x_1 x_2 x_4 \lor x_1 x_3 x_4$
$w_3 = x_1 x_2 x_3$	$w_7 = x_1 x_2 x_3 \lor x_1 x_2 x_4 \lor x_1 x_3 x_4 \lor x_2 x_3 x_4$
$w_4 = x_1 x_2 x_3 \lor x_1 x_2 x_4$	$w_8 = x_2 x_3 x_4 \lor x_1 x_3 x_4 \lor \overline{x}_1 x_2 x_3 \lor x_1 \overline{x}_2 x_4 \lor x_1 x_2 \overline{x}_3 \overline{x}_4$

Table 1.13: The number of W-classes for $n = 1, 2, 3, 4, 5$.

n	1	2	3	4	5
$|C_W|$	1	2	3	8	48

The Walsh spectrum for f in the ordering by Rademacher functions, with the scaling factor 2^{-4} omitted, is

$$\mathbf{W}_f = [-4, -8 - 0, 0, 8, 4, -4, -4, -4, 4, -4, 0, 0, 0, 0, -4]^T.$$

We want to move the coefficient of the largest positive value r_4 at the position of r_0. Thus, we perform the disjoint spectral translation $f_1 = f \oplus x_4$. Among all coefficients, if the subscript 4 appears, it is deleted. If it does not, it is appended. This produces the spectrum

$$\mathbf{W}_{f_1} = [8, -4, 4, -4, -4, 0, 0, -8, 0, 0, 0, -4, 4, -4, -4, 0]^T.$$

We now move the coefficient $r_{14} = -8$ at the position of the coefficient r_1. For this, we perform the spectral translation $x_1 = x_1 \oplus x_4$. Thus, in all the coefficients having the subscript 1, we delete the subscript 4 if it appears, and we append it if does not appear. This produces the spectrum

$$\mathbf{W}_{f_2} = [8, -8, 4, -4, -4, 4, -4, -4, 0, 0, 0, 0, 0, 0, -4, -4]^T.$$

Now, we change the sign of r_1, which corresponds to the negation of x_1, and requires that we change the sign of all the coefficients having the subscript 1. This produces the spectrum

$$\mathbf{W}_{f_3} = [8, 8, 4, -4, -4, -4, 4, 4, 0, 0, 0, 0, 0, 0, -4, 4]^T.$$

Then, we change the sign of r_3, which corresponds to the negation of x_3, i.e., we change the sign of all the coefficients having the subscript 3. This, produces the spectrum

$$\mathbf{W}_{f_4} = [8, 8, 4, 4, -4, -4, -4, 4, 0, 0, 0, 0, 0, 0, 4, -4]^T.$$

Finally, we change the sign of r_4, which corresponds to the negation of x_4, and requires a change in the sign of all the coefficients having the subscript 4. This, produces the spectrum

$$\mathbf{W}_{f_5} = [8, 8, 4, 4, 4, -4, -4, -4, 0, 0, 0, 0, 0, 0, -4, 4]^T.$$

This spectrum is the row 6 in Table 1.11, which is the Walsh spectrum for a function f_k defined by the truth-vector

$$\mathbf{F}_k = [0, 0, 0, 0, 0, 0, 0, 0, 0, 0, 0, 1, 0, 1, 1, 1]^T.$$

Therefore, the function f and f_k are in the same W-class. To verify, we perform the classification operations used. Since these operations can be performed in an arbitrary order, we first negate all the variables, which converts $\mathbf{F}_k$ into

$$\mathbf{F}_{\bar{k}} = [1, 0, 0, 0, 1, 1, 1, 0, 0, 0, 0, 0, 0, 0, 0, 0]^T.$$

The spectral translation $x_1 = x_1 \oplus x_4$, reorders $\mathbf{F}_{\bar{k}}$ as

$$\mathbf{F}_{\bar{k},14} = [1, 0, 0, 0, 1, 0, 1, 0, 0, 0, 0, 0, 0, 1, 0, 0]^T.$$

Finally, the disjoint spectral translation $f_{\bar{k}14} \oplus x_4$ produces the truth-vector of f.

In [23], it has been shown that SD-classification operations are equivalent to spectral invariant operations between the zero and first order Walsh coefficients (the coefficients whose index is $w = 2^k$, $k = 1, \ldots, n - 1$). Therefore, the classification by Walsh coefficients is considerably stronger than the SD-classification. For instance, if $n \leq 4$, there are 83 SD-classes compared to 8 classes in the classification by Walsh coefficients.

The classification of switching functions by disjoint spectral translation corresponds to the classification in terms of group theory and Fourier transforms on finite Abelian groups [33].

In [49], the method of synthesis of any switching function by a threshold logic gate and a set of EXOR gates has been related to the work in references [26], [27], [32], see also the reply of Edwards to the comment by C.K. Yuen [14]. This method is related to approximation of switching functions by Walsh series with less than 2^n coefficients, actually $n + m + 1$ coefficients, with m a small integer (called Chebyshev approximation in [26]).

1.7   CLASSIFICATION USING DECISION DIAGRAMS

The first ideas about classification of switching functions by referring to the *shape of decision diagrams* have been presented in [42]. These considerations were motivated by the fact that logic functions, when represented by decision diagrams, are easily mapped to circuits. The layout of the circuit is directly determined by the shape of the decision diagram [12], [48].

The shape of a decision diagram is specified by the distribution of nodes per levels and the interconnections of nodes in the diagram [40]. The distribution of nodes can be specified by a vector $D = [k_1, \ldots, k_{n+1}]$, where k_i, $i = 1, \ldots, n$, is the number of nodes at the level i by starting from the root node at the level $i = 1$. Thus, $k_1 = 1$ for any diagram, and k_{n+1} is the number of constant nodes. Interconnections in a diagram are specified as a list of nodes to which the given node is connected for each node.

For classification purposes, when referring to the LP-classification, it is sufficient to consider the distribution of nodes per levels and interconnections. We can also neglect concrete values of constant coefficients. However, we must take into account the number of different values since it affects the distribution of nodes in a diagram.

In this section, we show that some of the good features of the LP-classification and the classification in terms of Walsh coefficients can be combined and shared when LP-representative functions are represented by Walsh decision diagrams (WDDs) [43]. In this way, the LP-equivalence classes are described by WDDs with a characteristic distribution of nodes per levels. Due to exploiting the spectral translation together with LP-classification rules, the number of WDDs with different distribution of nodes required to represent LP-representative functions is smaller than the number of LP-classes.

Table 1.14 shows the truth vectors of these LP-representative for $n = 3$ and their Walsh spectra in $(1, -1)$ encoding. Fig. 1.2 shows WDDs for LP-representatives for $n = 3$. The minimal number of nodes in the corresponding diagrams is three and the maximal number is 6. Table 1.15 lists the valid node distributions. We can observe that the maximum of three WDDs share the same node distribution.

Table 1.14: Truth-vectors and Walsh spectra of LP-representatives for $n = 3$.

f	F	W_f
f_1	[00000000]	$[8, 0, 0, 0, 0, 0, 0, 0]$
f_2	[00000001]	$[6, 2, 2, -2, 2, -2, -2, 2]$
f_3	[00000110]	$[4, 0, 0, 4, 4, 0, 0, -4]$
f_4	[00010110]	$[2, 2, 2, 2, 2, 2, 2, -6]$
f_5	[00011000]	$[4, 0, 0, -4, 0, 4, 4, 0]$
f_6	[01101011]	$[-2, -2, 2, 2, 2, 2, -2, 6]$

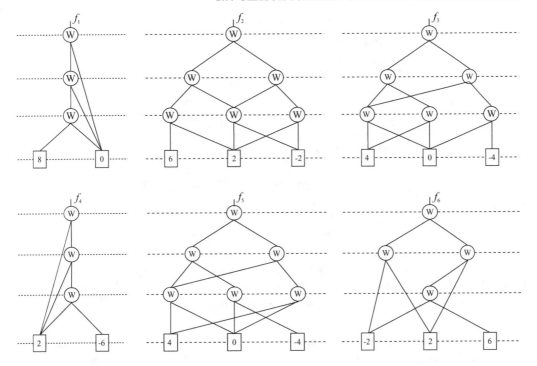

Figure 1.2: WDDs for LP-representatives for $n = 3$.

Table 1.15:	Node distributions in WDDs of	
LP-representatives for $n = 3$.		
Distribution	**# of functions**	**Functions**
$1, 1, 1, 2$	2	(f_1, f_4)
$1, 2, 3, 3$	3	(f_2, f_3, f_5)
$1, 2, 1, 3$	1	(f_6)

The introduction of spectral translation into LP-classification by representing LP-representatives by WDDs will be explained by the following example. It shows that by using LP-classification rules together with spectral translation (linear combination of variables in the Boolean domain) LP-representatives with the same distribution of nodes can be converted into forms that allows us to represent them by WDDs with identical interconnections, i.e., by WDDs with the same shape.

Example 1.9 The LP-representatives f_2, f_3 and f_5 for $n = 3$ can be represented by the WDDs with the same distribution of nodes and identical interconnections among them. The difference is

Table 1.16: Reordering of truth-vectors of f_3 and f_5 to be represented by the WDDs of the same shape as the WDD for $f_2, q = x_1 \oplus x_2$.

x_1, x_2, x_3	S_{f_2}	x_1, q, x_3	S_{f_3}	$x_1, \overline{q}, x_3$	S_{f_5}
000	6	000	4	010	0
001	2	001	0	011	−4
010	2	010	0	000	4
011	−2	011	4	001	0
100	2	110	0	100	0
101	−2	111	−4	101	4
110	−2	100	4	110	4
111	2	101	0	111	0

in labels at the edges. To achieve such representations, we reorder the elements of the Walsh spectra S_{f_3} and S_{f_5}. We perform the spectral translation $x_2 \to x_1 \oplus x_2$ and $x_2 \to \overline{x_1 \oplus x_2}$ for S_{f_3} and S_{f_5}, respectively. The values of the constant nodes are determined by an obvious encoding $6 \to -4$, $-2 \to 4$, and $2 \to 0$.

Fig. 1.2 shows WDDs for f_2, f_3 and f_5. Fig. 1.3 shows the corresponding Walsh decision trees with indicated permutation of labels at the edges, which has been performed to derive the corresponding WDDs.

It is interesting to notice that two functions f_4 and f_6, which are the most complex in terms of the number of terms in MSOPs [30] are not the most complex in terms of the number of nodes in WDDs.

Example 1.10   In Table 1.1, functions $f_1 = x_1 x_2 \vee x_3$ and $f_{10} = x_1 x_2 \vee x_2 x_3 \vee \overline{x}_2 \overline{x}_3$ belong to the same WDD-class for $n = 3$. This is the class defined by the distribution of nodes $[1, 2, 1, 3]$. The Walsh spectra of these functions are $\mathbf{W}_{f_1} = [-2, 6, 2, 2, 2, 2, -2, -2]^T$ and $\mathbf{W}_{f_{10}} = [-2, -2, 2, -6, 2, 2, -2, -2]^T$. We perform encoding of constant nodes as $-2, 2, 6 \to 2, -2, 6$. The labels at the edges, starting from the root node, and showing first left and then right edge for each node at each level, are as in the following table.

f_1	$1 - 2x_1$	1	$1 - 2x_2$	1	1	$1 - 2x_2$	1	$1 - 2x_3$
f_{10}	$1 - 2x_1$	1	1	$1 - 2(\overline{x_1 \oplus x_2})$	$1 - 2(\overline{x_1 \oplus x_2})$	1	1	$1 - 2x_3$

Examples 1.9 and 1.10 explain that when LP-representatives are represented by WDDs, we can perform permutation of labels at edges of certain selected nodes and encoding of values of constant nodes. Due to the recursive structure of decision trees, permutation of values for constant nodes corresponds to permutation of variables, while permutation of labels at the edges corresponds

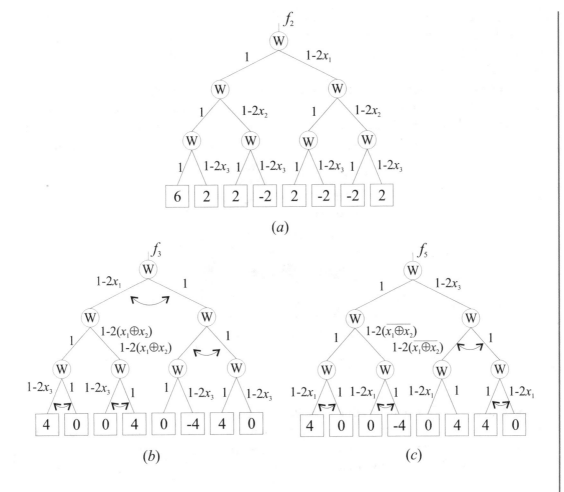

Figure 1.3: Walsh decision trees for f_2, f_3, f_5 and conversions among them.

to spectral translations. In this way, through WDDs, the set of LP-classification rules is enriched with the spectral translation, which is used as an operation in classification of switching functions by Walsh coefficients, see for instance [13], [22], [24], [44]. Due to that, it is possible to represent 6 LP-representatives in LP-classification for $n = 3$ by WDDs of three different shapes.

The same consideration can be extended to larger values of n in a straightforward manner.

For $n = 4$, there exist 14 different node distributions in WDDs. There are four distributions which are specific for a single function. Other distributions are shared by two or at most five functions. Functions that have the same distribution of nodes in WDDs can be reduced to each other by negation of variables and linear transformation of variables.

Table 1.17 shows the numbers of LP-representatives for $n = 3, 4, 5$ and the numbers of WDDs with different distribution of nodes to represent them. This table has been computed by examining all possible cases using the XML-environment for manipulating and calculating with decision diagrams [45].

For $n = 5$, the number of functions that can be represented by the same WDD under permutation of variables, spectral translation, and encoding of constant nodes, ranges from a single function (104 WDDs) to 402 functions for the WDD with the distribution of nodes per levels as $[1, 2, 4, 8, 11, 7]$. On average, one WDD can represent more than 18 LP-representatives for $n = 5$.

Table 1.17: Number of LP-representatives and WDDs with the same distribution of nodes to represent them.

n	$|C_{LP}|$	# of WDDs
3	6	3
4	30	14
5	6,936	383

The above considerations permit the following:

Conjecture 1.11   If distribution of nodes of Walsh decision diagrams for functions represented by two circuits are equal, the circuits are functionally equivalent under the permutation of labels at the edges, and encoding of values of constant nodes selected from prescribed sets of Walsh coefficients of switching functions for a given number of variables. In the Boolean domain, two functions

with equal distribution of nodes in WDDs are equivalent under LP-classification rules and linear transformation of variables. This linear transformation is determined by the spectral translation of Walsh coefficients.

1.8   CLOSING REMARKS

Partitioning of the set of all switching functions of a given number of variables into classes of functions that are similar to each other is useful in solving many problems on representations, realizations, and applications of switching functions. The chief idea is that, in a particular classification defined by a set of classification rules, all the functions in any class are represented by a single representative function. All other functions in the same class can be derived by applying classification rules. This immediately determines the extra circuitry needed to realize the particular function using the realization for the representative function.

The set of classification rules depends on the intended application and the technology. During the relatively short history of switching theory, classification rules have been formulated in different ways to serve some particular purposes, as well as have been adapted to various representations of switching functions. Further requirements are that the number of different classes should be reasonably small with the classification performed by a few simple classification rules. Attempts to fulfill these contradicting criteria have produced a variety of classification methods for switching functions, and in this chapter, we have reviewed a few of the most important ones.

Fig. 1.4 shows relationships among equivalence classes in different classifications discussed above. The LP-class does not allow the complementation of a function, and it is a refinement of the NP-class. WDD-class uses Walsh spectra of LP-representatives, and it is, therefore, related to both LP-class and W-class. If we further allow complementation of a function, which in the Walsh spectrum results in the change of the sign of all the coefficients, it can be related to the NPN-class.

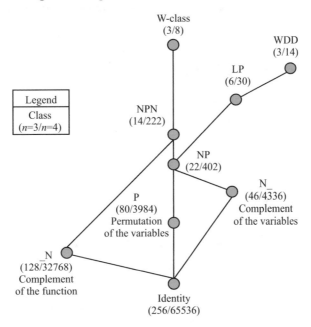

Figure 1.4: Relationships among equivalence classes.

ACKNOWLEDGMENTS

Authors are grateful to the editors and the reviewers whose thorough review and constructive comments were very useful to improve the presentation and discussions. This work was supported by the Academy of Finland, Finnish Center of Excellence Programme, Grant No. 213462.

1.9   EXERCISES

1.1.   Consider three functions: $f_1 = x_1x_2 \vee x_3$, $f_2 = x_1x_3 \vee x_2$, $f_3 = x_1\bar{x}_3 \vee x_2x_3 \vee \bar{x}_1x_2$. Check if these functions belong to the same P, NP or NPN-class.

1.2.   Determine all functions of $n \leq 2$ variables that belong to same NP and P class with the function $f(x_1, x_2) = x_1$.

1.3.   Determine all functions of $n \leq 3$ variables that belong to same P, NP and NPN-class with the function $f(x_1, x_2, x_3) = x_1 \vee x_2$.

1.4.   Determine examples of functions that belong to the same P, NP and NPN-class as a 5-variable function $f(x_1, x_2, x_3, x_4, x_5) = x_1x_2 + \bar{x}_1x_3 \vee x_4x_5 \vee x_3\bar{x}_5$.

1.5.   Calculate the number of functions that are NPN-equivalent to the function $f_1 = x_1 \vee x_2 \vee x_3$.

1.6.   Consider two NPN-equivalent functions: $f_1 = x_1x_2 \vee x_3$, and $f_2 = x_1\bar{x}_3 \vee x_2$. Compute W_p and W_l vectors of these functions and prove that they are indeed equivalent.

1.7.   Give an example of a self-dual function for $n = 3$.

1.8.   Present an example of a function which belongs to the same SD-class as the majority function of three variables $f(x_1, x_2, x_3) = x_1x_2 \vee x_2x_3 \vee x_1x_3$.

1.9.   Check if functions $f_1 = x_1 \vee x_2 \vee x_3$, and $f_2 = x_2(x_1 \vee x_3)$, belong to the same SD-class.

1.10.   Determine a function that belongs to the same LP-class as the function $f(x_1, x_2) = x_1 \vee \bar{x}_2$.

1.11.   Consider two switching functions: $f_1 = x_1 \vee x_2$, and $f_2 = \bar{x}_1 \oplus x_1x_2$. Are these functions LP-equivalent?

1.12.   Consider two N-equivalent functions: $f_1 = x_1x_2 \vee x_3$, $f_2 = \bar{x}_1x_2 \vee x_3$. Prove that these two functions are LP-equivalent by deriving the matrix $\mathbf{G}$ such that $\mathbf{F}_1 = \mathbf{G}\mathbf{F}_2$, where $\mathbf{F}_1$ and $\mathbf{F}_2$ are truth-vectors of f_1 and f_2, respectively.

1.13.   Consider two LP-equivalent functions: $f_1 = x_1x_2 \vee x_3$, $f_2 = \bar{x}_1\bar{x}_3 \vee x_2x_3$. In this case, we have a relation: $\mathbf{F}_1 = \mathbf{G}\mathbf{F}_2$, where $\mathbf{F}_1$ and $\mathbf{F}_2$ are truth vectors of f_1 and f_2, respectively. The matrix $\mathbf{G}$ is a Kronecker product representable matrix, with submatrices selected among (2×2) matrices defining the LP-operations. Demonstrate that these functions are not NPN-equivalent by examining the matrix $\mathbf{G}$.

1.14.   Present an example of a function which belongs to the same class as the function: $f_1(x_1, x_2, x_3, x_4) = [0001]^T$, according to the properties of its Walsh spectrum. The function f_1 is given in the form of a vector of hexadecimal values.

1.15. Construct and compare the distribution of nodes per levels in WDDs for functions $f_1(x_1, x_2, x_3, x_4) = [0168]$ and $f_2(x_1, x_2, x_3, x_4) = [016a]$. Functions are given in the form of vectors of hexadecimal values. Do these functions belong to the same class according to the properties of their Walsh spectra?

1.16. Consider a function $f(x_1, x_2, x_3) = \overline{x}_1 x_2 \vee \overline{x}_3$. Determine a self-dual function f^{sd} of four variables assigned to f. Show an example of a function that is in the same SD-class with f and f^{sd}.

1.17. Consider two W-equivalent functions: $f_1 = x_1 \overline{x}_2 x_3 x_4$, $f_2 = x_1 x_2 x_3 x_4$. The function f_2 can be derived from the function f_1 by the application of $x_2 \leftrightarrow x_1 \oplus x_2$ which is a Walsh invariant operation. Demonstrate that this operation indeed corresponds to a permutation of Walsh coefficients.

REFERENCES

[1] I.N. Aizenberg, N.N. Aizenberg, J. Vandewalle, *Multi-Valued and Universal Binary Neurons — Theory, Learning, Applications*, Kluwer Academic Publishers, Boston, Dordrecht, London, 2000. 1.1, 1.5

[2] N.N. Aizenberg, Yu.L. Ivaskiv, *Multivalued Threshold Logic*, Naukova Dumka, Kiev, Ukraine, 1977, in Russian. 1.1

[3] H. Astola, "Computing the LP- and NP-equivalence classes using Burnside's lemma," Department of Signal Processing, Tampere University of Technology, Research Seminar, July 2009. 1.3, 1.4

[4] J.T. Astola, R.S. Stanković, *Fundamentals of Switching Theory and Logic Design*, Springer, 2006. 1.3, 1.3, 1.3

[5] P.J. Cameron, *Combinatorics: Topics, Techniques, Algorithms*, Cambridge University Press, 1994. 1.3

[6] C.H. Chang, B.J. Falkowski, "*NPN*-classification using weight and literal vectors of Reed-Muller expansion," *Electronics Letters*, Vol. 35, No. 10. 1999, 798-799. 1.3

[7] C.H. Chang, B.J. Falkowski, "Reed-Muller weight and literal vectors for *NPN*-classification," *Proc. IEEE Int. Symp. on Circuits and Systems*, ISCAS '99, May 30 - June 2, 1999, Vol. 1, 379-382. DOI: 10.1109/ISCAS.1999.777882 1.3

[8] C.K. Chow, "On the characterization of threshold functions," *Proc. IEEE Symp. Switching Theory and Logic Design*, 1961, 34-38. 1.6.1, 1.6.2

[9] S. Dautović, L. Novak, "A comment on 'Boolean classification via fixed-polarity Reed-Muller Form'," *IEEE Trans. Computers*, Vol. C-55, No. 8, 2006, 1067-1069. DOI: 10.1109/TC.2006.114 1.3

[10] D. Debnath, T. Sasao, "Fast Boolean matching under permutation by efficient computation of canonical form," *IEICE Trans. Fundamentals*, Vol. E87-A, No. 12, 2004, 3134-3140. 1.3

[11] D. Debnath, T. Sasao, "Efficient computation of canonical form under variable permutation and negation for Boolean matching in large libraries," *IEICE Trans. Fundamentals*, Vol. E-89A, No. 12, 2006, 3443-3450. DOI: 10.1093/ietfec/e89-a.12.3443 1.3

[12] G. De Micheli, *Synthesis and Optimization of Digital Circuits*, McGraw Hill, 1994. 1.3, 1.7

[13] C.R. Edwards, "The application of the Rademacher-Walsh transform to Boolean function classification and threshold logic synthesis," *IEEE Trans. Computers*, Vol. C-24, No. 1, 1975, 48-62. DOI: 10.1109/T-C.1975.224082 1.6.1, 1.6.2, 1.7

[14] C.R. Edwards, "Author's Reply to the Comment by Yuen [50]," *IEEE Trans. Computers*, 1976, 767. 1.6.1, 1.6.2

[15] S.W. Golomb, "On the classification of Boolean functions," *IRE Trans. Circuit Theory*, Vol. CT-6, 1959, 176-186. DOI: 10.1109/TIT.1959.1057543 1.3

[16] E. Goto, "Threshold, majority and bilateral switching devices," in Aiken, H., Main, W.F., (eds.), *Switching Theory in Space Technology*, Proceedings of the Symposium held in Sunnyvale, California, February 27-28 and March 1, 1962, Stanford University Press, 1963. 1.5

[17] E. Goto, H. Takahasi, "Some theorems in threshold logic for enumerating Boolean functions," *Proc. IFIP Congress*, North-Holland, 1962, 747-752. 1.5, 1.5, 1.5

[18] T. Gowda, S. Leshner, S. Vrudhula, G. Konjevod, "Synthesis of threshold logic circuits using tree matching," *Proc.18th European Conference on Circuit Theory and Design*, 2007. DOI: 10.1109/ECCTD.2007.4529730 1.5

[19] T. Gowda, S. Vrudhula, "A decomposition based approach for synthesis of multi-level threshold logic circuits," *Proc. Asia and South Pacific Design Automation Conference*, Seoul, Korea, 21 January 2008, 125 - 130. 1.5

[20] M.A. Harrison, *Introduction to Switching and Automata Theory*, McGraw-Hill, 1965. 1.3, 1.3, 1.3

[21] M.A. Harrison, *Combinatorial Problems in Boolean Algebras and Applications to the Theory of Switching*, Doctoral Thesis, University of Michigan, USA, 1962. 1.3, 1.3

[22] D.L. Hurst, *Logical Processing of Digital Signals*, Crane Russak and Edward Arnold, London and Basel, 1978. 1.3, 1.6.1, 1.6.1, 1.6.2, 1.6.2, 1.6.2, 1.7, A.1

[23] S.L. Hurst, "The relationship between the self-dualized classification of Boolean functions and spectral coefficient classification," *Int. Journal of Electronics*, Vol. 56, No. 6, 1984, 801-808. 1.6.2

[24] S.L. Hurst, D.M. Miller, J.C. Muzio, *Spectral Techniques in Digital Logic*, Academic Press, Bristol, 1985. 1.3, 1.6.2, 1.7

[25] A. Jain, R. Bryant, "Inverter minimization in logic networks," *Proc. Int. Workshop on Logic Synthesis*, 1993, 9c1-9c6. 1.3

[26] K.R. Kaplan, R.O. Winder, "Chebyshev approximation and threshold functions," *IEEE Trans. Electron. Comput.*, (Short Notes), Vol. EC-14, 1965, 250-252. DOI: 10.1109/PGEC.1965.264254 1.6.2

[27] R. Kascerman, "A linear summation threshold device," *IEEE Trans. Electron. Comput.*, (Corresp.), Vol. EC-12, 1963, 914-915. DOI: 10.1109/PGEC.1963.263597 1.6.2

[28] N. Koda, T. Sasao, "LP-equivalence class of logic functions," *IFIP 10.5 Workshop on Application of the Reed-Muller expansion in Circuit Design*, Hamburg, Germany, September 1993, 99-106. 1.5, 1.3, 1.4, 1.4

[29] N. Koda, T. Sasao, "LP-Characteristic vectors of logic functions and their applications," *Trans. IEICE Japan*, Part D-I, Vol. J76-D-1, No. 6, 1993, 260-268, (in Japanese). 1.4

[30] N. Koda, T. Sasao, "An upper bound on the number of products in minimum ESOPs," *IFIP WG 10.5 Workshop on Applications of the Reed-Muller Expansions in Circuit Design* (Reed-Muller 95), Makuhari, Japan, August 1995, 27-29. 1.4, 1.4, 1.9

[31] N. Koda, T. Sasao, "A method to simplify multiple-output AND-EXOR expressions" (in Japanese), *Trans. IEICE*, Vol. J79-D-1, No. 2, 1996, 43-52. DOI: 10.1002/scj.4690270901 1.4, 1.4

[32] C.L. Lawson, *Contributions to the Theory of Linear Least Maximum Approximation*, Ph.D. dissertation, Univ. California, Los Angeles, 1961. 1.6.2

[33] R.J. Lechner, "Harmonic analysis of switching functions," in Mukhopadhyay, A., (ed.), *Recent Developments in Switching Theory*, Academic Press, 1971, 122-228. 1.6.2

[34] C. Moraga, "Introducing disjoint spectral translation in spectral multiple-valued logic design," *Electronics Letters*, Vol. 14, No. 8, 1978, 241-243. DOI: 10.1049/el:19780164 1.6.2

[35] S. Muroga, *Logic Design and Switching Theory*, John Wiley and Sons, 1979, Reprinted edition Krieger Publishing Company, Malaber, FL, 1990. 1.1, 1.3, 1.3

[36] T. Sasao, "A transformation of multiple-valued input two-valued output functions and its application to simplification of exclusive-or sum-of-products expressions," *Proc. Int. Symp. on Multiple-Valued Logic*, Victoria, British Columbia, Canada, May 26-29, 1991, 270-279. 1.4

[37] T. Sasao, *Switching Theory for Logic Synthesis*, Kluwer Academic Publishers, 1999. 1.1, 1, 1.3, 1.3, 1.3

[38] T. Sasao, Ph.W. Besslich, "On the complexity of mod 2 PLA's," *IEEE Trans. Comput.*, Vol. C-39, No. 2, 1991, 262-266. DOI: 10.1109/12.45212 1.4

[39] T. Sasao, M. Fujita, (eds.), *Representations of Discrete Functions*, Kluwer Academic Publishers, 1996. 43

[40] R.S. Stanković, "Some remarks on basic characteristics of decision diagrams," *Proc. 4th Int. Workshop on Applications of Reed-Muller Expansion in Circuit Design*, August 20-21, 1999, Victoria, B.C., Canada, 139-146. 1.7

[41] R.S. Stanković, J.T. Astola, B. Steinbach, "Former and recent work in classification of switching functions," *Proc. 8th Int. Workshop on Boolean Problems*, Freiberg, Germany, September 18-19, 2008. 1.3

[42] R. S. Stanković, C. Moraga, T. Sasao, "Spectral transform decision diagrams," *Proc. IFIP WG 10.5 Workshop on Application of the Reed-Muller Expansion in Circuit Design, Reed-Muller'95*, August 27-29, 1995, Makuhari, Chiba, Japan, 46-53. 1.7

[43] R.S. Stanković, T. Sasao, C. Moraga, "Spectral transform decision diagrams" in [39], 55-92. 1.7

[44] R.S. Stanković, S. Stanković, J.T. Astola, "Remarks on the representation of prototype functions in LP-classification by Walsh decision diagrams," *Proc. Reed-Muller Workshop*, Naha, Okinawa, Japan, May 23-24, 2009. 1.7

[45] S. Stanković, J. Astola, "XML framework for various types of decision diagrams for discrete functions," *IEICE Trans, Information & Systems*, Vol.E90-D, No.11, 1731-1740. DOI: 10.1093/ietisy/e90-d.11.1731 1.7

[46] C.-C. Tsai, M. Marek-Sadowska, "Boolean function classification via fixed-polarity Reed-Muller forms," *IEEE Trans. Computers*, Vol. C-46, No. 2, 1997, 173-186. DOI: 10.1109/12.565592 1.3

[47] J.H. Van Lint, R.M. Wilson, *A Course in Combinatorics*, Cambridge Univ. Press, 1992. 1.3

[48] S.N. Yanushkevich, D.M. Miller, V.P. Shmerko, R.S. Stanković, *Decision Diagram Technique for Micro - and Nanoelectronic Design*, CRC Press, Taylor & Francis, Boca Raton, London, New York, 2006. 1.7

[49] C.K. Yuen, "Function approximation by Walsh series," *IEEE Trans. Computers*, Vol. C-24, 1975, 590-598. DOI: 10.1109/T-C.1975.224271 1.6.2

[50] C.K. Yuen, "Comments on 'The application of the Rademacher-Walsh transform to Boolean function classification and threshold logic synthesis'," *IEEE Trans. Computers*, July 1976, 766-767. DOI: 10.1109/TC.1976.1674688 14

[51] R. Zhang, P. Gupta, L. Zhong, N.K. Jha, "Synthesis and optimization of threshold logic networks with application to nanotechnologies," in Lauwereins, R., Madsen, J., *Design, Automation, and Test in Europe The Most Influential Papers of 10 Years DATE*, Springer 2008, 325-346. 1.5

CHAPTER 2

Boolean Functions for Cryptography

Jon T. Butler and Tsutomu Sasao

CHAPTER SUMMARY

Significant research has been done on bent functions, yet researchers in switching theory have paid little attention to this important topic. The goal of this paper is to provide a concise exposition. Bent functions are the most nonlinear functions among n-variable switching functions, and are useful in cryptographic applications. This chapter discusses three other cryptographic properties - strict avalanche criterion, propagation criterion, and correlation immunity. We discuss known properties, as well as open questions. It is assumed that the reader is familiar with switching circuit theory. Familiarity with Reed-Muller expansions is helpful, but not essential.

2.1 INTRODUCTION

One approach to encoding a plaintext message into cyphertext is to use one 7 bit key for each 7 bit ASCII character and to apply the bitwise exclusive-OR to each letter. In this way, each letter of the plaintext message is converted to a different letter in the cyphertext. Decryption is simple. Just apply the same key to the cyphertext. Since the second application of the key "annihilates" the first application, we are left with the plaintext letter. The problem with this is that the distribution of probabilities of the letters in the plaintext also occurs in the cyphertext. This can be exploited by someone listening to the cyphertext. For example, the most frequent letters in the cyphertext may be assumed to be "e" or "t" and the least frequent letters may be assumed to be "z" or "q".

To avoid decryption by an outsider, one seeks a key stream that is random. However, high-speed, parallel computers can be used to exploit variations from randomness in the key stream. For example, in a "linear" attack, a key stream is tried that is generated from a linear Boolean function. If the actual key stream used in encryption is close to linear, there will be errors, but such an attack may be ultimately successful. Against such attacks, one seeks a function that is as far from linear as possible. These are the bent functions.

In the next section, we introduce bent functions and discuss their properties. In the third section, we discuss symmetric bent functions. Then, in the next three sections, we discuss three classes of functions that have other cryptographic properties. These are the strict avalanche criterion, the propagation criterion, and the correlation immunity. Then, we provide concluding remarks.

2.2   PROPERTIES OF BENT FUNCTIONS

The term **bent function** describes functions that are the "most nonlinear" of the n-variable functions. It was introduced in 1976 by **Rothaus** [17]. Presumably, "bent" was chosen since it is an antonym of "linear". Rothaus' seminal work [17] was actually completed ten years earlier, but remained under restricted circulation until 1976. Rothaus died in 2003, six days before he was scheduled to retire from the Department Mathematics at Cornell University. His work was chosen for inclusion in Knuth's long-anticipated "The Art of Computer Programming, Volume 4" [9].

Definition 2.1   A **linear function** is the constant 0 function or the exclusive-OR of one or more variables.

Example 2.2   There are eight 3-variable linear functions, $0, x_1, x_2, x_3, x_1 \oplus x_2, x_1 \oplus x_3, x_2 \oplus x_3$, and $x_1 \oplus x_2 \oplus x_3$. Only one of the eight functions actually depends on all three variables. However, because it simplifies the counting of functions, we will view all eight functions as functions of three variables.

Definition 2.3   An **affine function** is a linear function or the complement of a linear function[1].

Example 2.4   There are 16 different 3-variable affine functions, $0, x_1, x_2, x_3, x_1 \oplus x_2, x_1 \oplus x_3, x_2 \oplus x_3, x_1 \oplus x_2 \oplus x_3, 1, x_1 \oplus 1, x_2 \oplus 1, x_3 \oplus 1, x_1 \oplus x_2 \oplus 1, x_1 \oplus x_3 \oplus 1, x_2 \oplus x_3 \oplus 1, x_1 \oplus x_2 \oplus x_3 \oplus 1$.

Affine functions are one extreme type of switching function. We are interested in the extent to which a switching function departs from affine functions.

Definition 2.5   The **nonlinearity** NL_f of a function f is the minimum number of truth table entries that must be changed in order to convert f to an affine function.

---

[1]In papers on switching theory, the term "linear" is often used to describe an affine function. We adopt the terminology of Cusick and Stanica [2](p.6).

The nonlinearity of a function f is the minimum Hamming distance between the truth tables of f and an affine function[2].

Example 2.6   Among 3-variable functions, the function $f = x_1x_2x_3$, which is not affine, has nonlinearity 1, since converting the single 1 in its truth table to a 0 creates the constant 0 function, which is affine.

Definition 2.7   Let f be a Boolean function on n-variables, where n is even. f is a **bent function** if its nonlinearity is as large as possible.

Bent functions have the property that they are a maximum distance from all affine functions. For example, $f = x_1x_2 \oplus x_3x_4$ is a known bent function on 4 variables; it is a distance 6 from 16 of the 32 affine functions on 4 variables and a distance 10 from the other 16 affine functions. That is, at least six entries of the truth table of f must be changed to convert it into an affine function. Since there are no 4-variable functions whose minimal distance to an affine function is seven or larger, it follows that $f = x_1x_2 \oplus x_3x_4$ is bent.

Meier and Staffelbach [14] were the first to recognize that the set of functions with maximum nonlinearity was identical to the set of bent functions, as defined by Rothaus [17]. They showed that the maximum nonlinearity was $2^{n-1} - 2^{\frac{n}{2}-1}$.

Bent functions are important because of a cryptanalysis technique in which nonlinear functions used in the encryption process are approximated by linear functions. That is, when the encryption is linear, decryption is straightforward. When the encryption is "slightly nonlinear", then a linear approximation can be used, with an understanding that decryption is erroneous but potentially correctable. Indeed, Matsui [12, 13] proposes a linear attack of the Data Encryption Standard (DES). Bent functions are valued because they are the most difficult to approximate by linear functions.

Fig. 2.1 shows the distribution of nonlinearity values for all 65,536 functions on 4 variables. For example, Fig. 2.1 shows that 32, 512, 3840, and 17920 4-variable functions have a nonlinearity of 0, 1, 2, and 3, respectively. We expect 32 functions to have a nonlinearity of 0 because that is the number of affine 4-variable functions. The number of functions with nonlinearity 1 is 512. As it turns out, 512 is an upper bound on the number of functions with that nonlinearity. That is, for each 4-variable function with nonlinearity 0, there can be no more than 16 functions that are a Hamming distance 1 from it, for a total of $32 \times 16 = 512$ functions. It must be that, among the functions with nonlinearity 1, none are a Hamming distance 1 away from *two* or more affine functions.

A similar statement is true of 4-variable functions with nonlinearity 2 and 3. If all such functions are unique, then there are $\binom{16}{2}32 = 3,840$ and $\binom{16}{3}32 = 17,920$ functions respectively. As can be seen from Fig. 2.1, there are exactly 3,840 and 17,920 functions with nonlinearity 2 and 3, respectively.

---

[2] We note an inconsistency in the terminology. The term "nonaffinity" would be a more consistent alternative to "nonlinearity". However, most papers use the latter term.

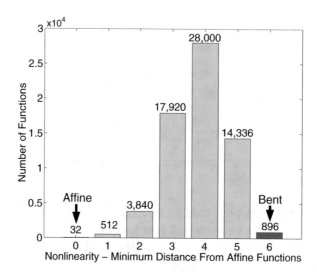

Figure 2.1: Distribution of all 4 variables to nonlinearity.

It follows that all functions with nonlinearities 0, 1, 2, and 3 are a minimum distance from exactly one affine function. For functions with nonlinearity 4 or more, the same statement is not true; for such functions, two or more affine functions are at the same minimum distance.

Fig. 2.1 shows that most 4-variable functions have a nonlinearity value near the middle, around 3, 4, and 5. By comparison, functions with extreme nonlinearity values, 0 and 6, are rare. Indeed, the fraction of all n-variable affine functions approaches 0 as n increases. Specifically, the fraction of functions that are affine, $2^{n+1}/2^{2^n}$, rapidly approaches 0 as $n \to \infty$. The extreme values are important. The 32 functions with nonlinearity 0 are the affine functions. There are 896 functions with nonlinearity 6; these are the bent functions. The exact number of bent functions is known only for small values of n. That is, an exact number of bent functions is known only for $n \leq 8$ [1, 10, 15]. The number of n-variable bent functions, for general n, is an open question that has resulted in a number of studies on bounds [1, 17]. Table 2.1 shows the number of bent functions for $2 \leq n \leq 8$. Note that, while the number of bent functions increases rapidly with increasing n, the proportion of functions that are bent decreases rapidly.

Definition 2.8   The **weight** $|f|$ of a function f is the number of 1's in the truth table of f.

Fig. 2.2 shows the distribution of 4-variable functions to the weight of the function and its nonlinearity. Specifically, a function contributes 1 to the count of functions that have a specified weight and a specified nonlinearity, NL_f. The vertical axis shows the *log* of the number of functions (to allow the display of both small and large values). There are seven graphs, one for each value of $NL_f = 0, 1, 2, 3, 4, 5,$ and 6. For example, the top graph shows the distribution of affine functions

Table 2.1: The number of n-variable bent functions for $2 \leq n \leq 8$.

n	# of Bent Functions	Fraction That Are Bent
2	$8 = 2^3$	2^{-1}
4	$896 = 2^{9.8}$	$2^{-6.2}$
6	$5{,}425{,}430{,}528 \approx 2^{32.3}$	$2^{-31.7}$
8	$\approx 2^{106.3}$	$2^{-149.7}$

with respect to weight for $NL_f = 0$. In this case, there is one function with weight 0 (the constant 0 function), 30 functions with weight 8, and one function with weight 16 (the constant 1 function). Interestingly, the distribution of 896 bent functions, as shown in the last graph, is simple. Specifically, 448 have weight 6 and 448 have weight 10. In general,

Theorem 2.9   [14] The weight of an n-variable bent function is $2^{n-1} \pm 2^{\frac{n}{2}-1}$.

Note that the bar chart in Fig. 2.2 is symmetric with respect to the center line of weight 8. This is because f and its complement, $\bar{f} = f \oplus 1$, have the same nonlinearity.

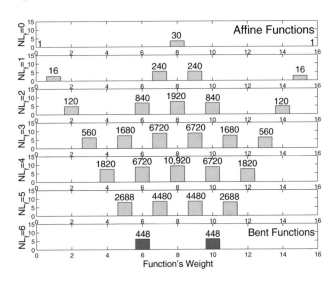

Figure 2.2: Distribution over 4-variable functions to nonlinearity and to the weight. (The *log* of the number of functions is plotted along the vertical axis).

We observed in Fig. 2.1 that all functions with nonlinearity 1 were a Hamming distance 1 from a unique function with nonlinearity 0. This can be seen in Fig. 2.2. For example, for the trivial affine function whose truth table is all 0's, there are 16 functions with $NL_f = 1$ that with weight 1. Similarly, for the trivial affine function whose truth table is all 1's, there are 16 functions with $NL_f = 1$ that have weight 15. For each of the 30 affine functions with weight 8, there are 16 functions that are a distance 1 away. This is shown by two bars each of height $30 \times 16/2 = 240$, one with weight 7 and the other with weight 9.

Definition 2.10   Switching function f is **NPN equivalent** to h iff f can be obtained from h by a complementation of variables (N), a permutation of variables (P), and a complementation of the function (N).

The following four results from Cusick and Stanica [2] relate bent functions to the NPN equivalence classes.

Lemma 2.11 f is a bent function iff $1 \oplus f$ is a bent function.

Lemma 2.12 $f(x_1, x_2, \ldots, x_i, \ldots, x_n)$ is a bent function iff $f(x_1, x_2, \ldots, \bar{x}_i, \ldots, x_n)$ is a bent function.

Lemma 2.13 $f(x_1, x_2, \ldots, x_i, \ldots, x_j, \ldots, x_n)$ is a bent function iff $f(x_1, x_2, \ldots, x_j, \ldots, x_i, \ldots, x_n)$ is a bent function.

From Lemmas 2.11, 2.12, and 2.13, we have

Theorem 2.14 f is a bent function iff any function in the same NPN equivalence class as f is a bent function[3].

Theorem 2.14 states that either all functions in an NPN equivalence class are bent or all are not bent. It follows that one way to enumerate bent functions is to enumerate the bent NPN equivalence classes. NPN equivalent classes are somewhat hard to characterize. They do not all contain the same number of functions. For example, symmetric functions tend to occur in small classes. However, for one equivalence relation, the equivalence classes all have the same size.

Definition 2.15   Switching function f is **A equivalent** to h iff $h = f \oplus g$, where g is any affine function. f and h are said to belong to the same **A-class**.

---

[3]This can be extended to *Theorem: All functions in an NPN equivalence class have the same nonlinearity.*

Note that if f is bent, then it is a "bent distance" from any affine function a. Then, consider a special affine function s. Since $s \oplus a$ is an affine function, f is a bent distance from $s \oplus a$, and it follows that $f \oplus s$ is a bent distance from a, and indeed from any affine function. This proves the following.

Lemma 2.16 f is a bent function iff $f \oplus g_{\text{affine}}$ is a bent function, where g_{affine} is an affine function.

Lemma 2.16 implies that functions in the same A-class are either all bent or all not bent.

Example 2.17   From Fig. 2.2, there are 896 bent functions on four variables. These are divided into equivalence classes with respect to the affine functions. Since there are 32 affine functions, there are $896/32 = 28$ equivalence classes. Note that, unlike NPN equivalence classes, these equivalence classes have the *same* number of elements, 2^{n+1}, as the number of affine functions.

The Walsh transform is useful in understanding the properties of bent functions. Indeed, Rothaus [17] originally defined bent functions to be switching functions whose Walsh transform contains only the values $\pm 2^{n/2}$. No switching function with odd n satisfies Rothaus' definition. The term "semi-bent" is often applied to switching functions whose Walsh transform contains only the values $\{0, \pm 2^{(n+1)/2}\}$, when n is odd. For example, the majority function on 3-variables, which is 1 if and only if 2 or 3 variables are 1, has a Walsh transform of $(0, 4, 4, 0, 4, 0, 0, -4)$. It is a semi-bent function.

Definition 2.18   A logic function is a mapping $\{0, 1\}^n \rightarrow \{0, 1\}$. When we consider the *Walsh transform of a logic function f, we often use the encoding $(1 - 2x_i)$ and $(1 - 2f)$, where x_i and f are viewed as integers i.e., 0 maps to 1 and 1 maps to -1). In this case, the function represents a mapping $\{1, -1\}^n \rightarrow \{1, -1\}$. Such an encoding is the $(1, -1)$ encoding, and the function expressed in this encoding is denoted as $\widehat{f}$.*

Definition 2.19   The **Walsh transform matrix** of n variables is

$$W_n = \begin{bmatrix} W_{n-1} & W_{n-1} \\ W_{n-1} & -W_{n-1} \end{bmatrix},$$

where

$$W_1 = \begin{bmatrix} 1 & 1 \\ 1 & -1 \end{bmatrix},$$

Let $\widehat{f}$ be the $(1, -1)$ encoding of the logic function f. Then, $\vec{s} = W_n \widehat{f}$ is the **Walsh spectrum** of f.

Consider $f = x_1 x_2 \oplus x_3 x_4$. The Walsh spectrum of f is calculated as follows. First, obtain the truth table and function table for $f = x_1 x_2 \oplus x_3 x_4$. The truth table and function table are shown in Table 2.2.

x_4	x_3	x_2	x_1	f	$1-2f$
0	0	0	0	0	1
0	0	0	1	0	1
0	0	1	0	0	1
0	0	1	1	1	-1
0	1	0	0	0	1
0	1	0	1	0	1
0	1	1	0	0	1
0	1	1	1	1	-1
1	0	0	0	0	1
1	0	0	1	0	1
1	0	1	0	0	1
1	0	1	1	1	-1
1	1	0	0	1	-1
1	1	0	1	1	-1
1	1	1	0	1	-1
1	1	1	1	0	1

$$W_4 \widehat{f} =$$

$$
\begin{bmatrix}
1 & 1 & 1 & 1 & 1 & 1 & 1 & 1 & 1 & 1 & 1 & 1 & 1 & 1 & 1 & 1 \\
1 & -1 & 1 & -1 & 1 & -1 & 1 & -1 & 1 & -1 & 1 & -1 & 1 & -1 & 1 & -1 \\
1 & 1 & -1 & -1 & 1 & 1 & -1 & -1 & 1 & 1 & -1 & -1 & 1 & 1 & -1 & -1 \\
1 & -1 & -1 & 1 & 1 & -1 & -1 & 1 & 1 & -1 & -1 & 1 & 1 & -1 & -1 & 1 \\
1 & 1 & 1 & 1 & -1 & -1 & -1 & -1 & 1 & 1 & 1 & 1 & -1 & -1 & -1 & -1 \\
1 & -1 & 1 & -1 & -1 & 1 & -1 & 1 & 1 & -1 & 1 & -1 & -1 & 1 & -1 & 1 \\
1 & 1 & -1 & -1 & -1 & -1 & 1 & 1 & 1 & 1 & -1 & -1 & -1 & -1 & 1 & 1 \\
1 & -1 & -1 & 1 & -1 & 1 & 1 & -1 & 1 & -1 & -1 & 1 & -1 & 1 & 1 & -1 \\
1 & 1 & 1 & 1 & 1 & 1 & 1 & 1 & -1 & -1 & -1 & -1 & -1 & -1 & -1 & -1 \\
1 & -1 & 1 & -1 & 1 & -1 & 1 & -1 & -1 & 1 & -1 & 1 & -1 & 1 & -1 & 1 \\
1 & 1 & -1 & -1 & 1 & 1 & -1 & -1 & -1 & -1 & 1 & 1 & -1 & -1 & 1 & 1 \\
1 & -1 & -1 & 1 & 1 & -1 & -1 & 1 & -1 & 1 & 1 & -1 & -1 & 1 & 1 & -1 \\
1 & 1 & 1 & 1 & -1 & -1 & -1 & -1 & -1 & -1 & -1 & -1 & 1 & 1 & 1 & 1 \\
1 & -1 & 1 & -1 & -1 & 1 & -1 & 1 & -1 & 1 & -1 & 1 & 1 & -1 & 1 & -1 \\
1 & 1 & -1 & -1 & -1 & -1 & 1 & 1 & -1 & -1 & 1 & 1 & 1 & 1 & -1 & -1 \\
1 & -1 & -1 & 1 & -1 & 1 & 1 & -1 & -1 & 1 & 1 & -1 & 1 & -1 & -1 & 1
\end{bmatrix}
\cdot
\begin{bmatrix}
1 \\ 1 \\ 1 \\ -1 \\ 1 \\ 1 \\ 1 \\ -1 \\ 1 \\ 1 \\ 1 \\ -1 \\ -1 \\ -1 \\ -1 \\ 1
\end{bmatrix}
=
\begin{bmatrix}
4 \\ 4 \\ 4 \\ -4 \\ 4 \\ 4 \\ 4 \\ -4 \\ 4 \\ 4 \\ 4 \\ -4 \\ -4 \\ -4 \\ -4 \\ 4
\end{bmatrix}.
$$

From this, we see that the coefficients of the Walsh spectrum for $f = x_1 x_2 \oplus x_3 x_4$ are either 4 or -4. Because the magnitudes of all coefficients are 4, we say the spectrum is **flat**. This is a general result. It is known that the Walsh spectrum of a bent function is flat with values $\pm 2^{\frac{n}{2}}$. The following interesting relation between the Walsh coefficients and nonlinearity is known.

Theorem 2.20   ([2], p. 13) The nonlinearity of a function f is

$$NL_f = 2^{n-1} - \frac{1}{2} max |W_n \widehat{f}|,$$

where $max|W_n\widehat{f}|$ is the maximum of the absolute values of the elements in $\vec{s} = W_n\widehat{f}$.

From Theorem 2.20, for $f = x_1x_2 \oplus x_3x_4$, $NL_f = 2^{4-1} - \frac{1}{2}4 = 6$.

Definition 2.21   The **PPRM (positive polarity Reed-Muller form)** of a function f is

$$f = a_0 \oplus a_1x_1 \oplus a_2x_2 \oplus \ldots \oplus a_nx_n \oplus a_{1,2}x_1x_2$$
$$\oplus a_{1,3}x_1x_3 \oplus \ldots \oplus a_{n-1,n}x_{n-1}x_n \oplus \ldots$$
$$\oplus a_{1,2,\ldots,a_n}x_1x_2\ldots x_n.$$

The PPRM of a function f is also called the **algebraic normal form** (ANF) of f (e.g. [2]).

The PPRM of a function f is unique.

Definition 2.22   The **degree of a product term** is the number of variables in that term. The **degree of a function** f is the maximum of the degrees among the product terms in the PPRM of f.

Lemma 2.16 implies that, given the PPRM of any bent function f, another bent function is realized by simply changing the coefficients of the constant or linear terms in the PPRM of f. One can take as the *representative* of the A-class of a bent function, the function whose constant and linear terms are all absent.

In the case of all 4-variable bent functions, it is known that the largest degree is 2. That is, in 4-variable *bent* functions, there are no terms in the PPRM with degree larger than 2. Further, at least one term of degree 2 is needed; otherwise, the function is affine. However, if a function is bent, permuting variables yields a bent function. It follows, for example, that, if a bent function, such as $x_1x_2 \oplus x_3x_4$, has two quadratic terms, then there is a bent function with quadratic terms $x_1x_3 \oplus x_2x_4$ and another bent function with quadratic terms $x_1x_4 \oplus x_2x_3$. In the second two functions, the variables are a permutation of the variables in the first function.

Fig. 2.3 shows all of the ways pairs of variables can be arranged in 4-variable functions. There are 11 ways pairs can occur, including the case where there are no pairs (shown at the very top). For each of these ways, there is a graph in Fig. 2.3 and an example function.

In all, there are $2^6 = 64$ ways possible choices for A-classes for 4-variable bent functions. However, from a previous discussion, we know that there are actually only 28 A-classes. The 5-th column from the left in Fig. 2.3 shows which functions are bent. Specifically, there are four sets involving 2, 3, 4, and 6 pairs of variables. One of the sets in Fig. 2.3 has exactly two pairs of variables such that no variable appears in more than one pair i.e., the pairs are disjoint). The "3" shown in the third column of the row labeled "DQF" ("Disjoint quadratic functions") means that there are three functions. These are $f = x_1x_2 \oplus x_3x_4$, $g = x_1x_3 \oplus x_2x_4$, and $h = x_1x_4 \oplus x_2x_3$. These functions have special significance.

Definition 2.23   The **disjoint quadratic function** or **DQF** [24] is

$$f = x_1x_2 \oplus x_3x_4 \oplus \ldots \oplus x_{n-1}x_n, \tag{2.1}$$

# of Edges	Symmetry Class	# Funcs. In This Class	Example	Bent	Symmetric	Special
0	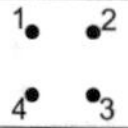	1	0		✓	Affine
1	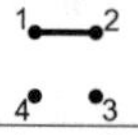	6	$x_1 x_2$			
2	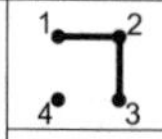	12	$x_1 x_2 \oplus x_2 x_3$			
2	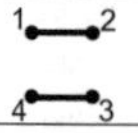	3	$x_1 x_2 \oplus x_3 x_4$	✓		DQF
3	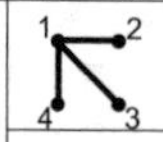	4	$x_1 x_2 \oplus x_1 x_3 \oplus x_1 x_4$			
3	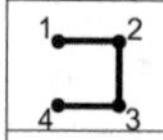	12	$x_1 x_2 \oplus x_2 x_3 \oplus x_3 x_4$	✓		
3	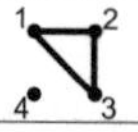	4	$x_1 x_2 \oplus x_2 x_3 \oplus x_1 x_3$			
4	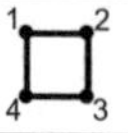	3	$x_1 x_2 \oplus x_2 x_3 \oplus x_3 x_4 \oplus x_1 x_4$			
4		12	$x_1 x_3 \oplus x_2 x_3 \oplus x_3 x_4 \oplus x_1 x_4$	✓		
5	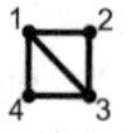	6	$x_1 x_2 \oplus x_1 x_3 \oplus x_2 x_3 \oplus x_3 x_4 \oplus x_1 x_4$			
6	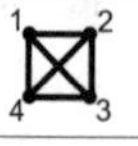	1	$x_1 x_2 \oplus x_1 x_3 \oplus x_2 x_3 \oplus x_2 x_4 \oplus x_3 x_4 \oplus x_1 x_4$	✓	✓	CQF
Total		64		28	2	

Figure 2.3: All arrangements of pairs of variables in 4-variable functions (Bent and symmetric functions are shown by check marks).

where n is an even positive integer.

This is similar to the Achilles' heel function, which has been defined using $\vee$ instead of $\oplus$ [19, 20]. It has often been offered as an example of how important variable order is in the realization of a function by a binary decision diagram (BDD). The disjoint quadratic function was among the first forms known to be bent [17].

Definition 2.24 The **complete quadratic function** or **CQF** is

$$f = x_1 x_2 \oplus x_1 x_3 \oplus \ldots \oplus x_i x_j \oplus \ldots \oplus x_{n-1} x_n, \tag{2.2}$$

where n is an even positive integer and $1 \leq i, j \leq n$, for $i < j$.

As can be seen in Fig. 2.3, this is the only function that is both bent and symmetric. We consider the CQF in the next section.

The observation that 4-variable bent functions have degree at most 2 can be extended. From Rothaus [17], the following surprising result is known.

Theorem 2.25   For $n > 2$, an n-variable bent function has degree at most $\frac{n}{2}$.

For $n = 2$, the degree of a bent function is 2. This represents a strong confinement on where a search for bent functions may be restricted. Rothaus [17] further showed that there exist bent functions on every degree d, where $2 \leq d \leq \frac{n}{2}$. There is significant interest in homogeneous bent functions [16, 23, 26].

Definition 2.26   A **homogeneous function** is a function whose PPRM consists of product terms all with the same degree.

Example 2.27   The disjoint quadratic function, $f = x_1x_2 \oplus x_3x_4 \oplus \ldots \oplus x_{n-1}x_n$, is homogeneous.

Xia, Seberry, Pieprzyk, and Charnes [26] showed the following.

Theorem 2.28   When $n > 6$, no n-variable homogeneous bent function has degree $\frac{n}{2}$.

Therefore, from [17] and [26], for $n > 6$, degree-$\frac{n}{2}$ n-variable bent functions exist, but none are homogeneous. The 4-variable disjoint quadratic function is an example of a 4-variable homogeneous bent function (of degree 2). Xia, Seberry, and Pieprzyk [16] show the existence of homogeneous 6-variable bent functions of degree 3. Indeed, 30 such functions exist. Thus, Theorem 2.28 does not hold for $n \leq 6$.

2.3   PROPERTIES OF SYMMETRIC BENT FUNCTIONS

Definition 2.29   A **symmetric function** is unchanged by any permutation of its variables. Regarding symmetric functions, in 1994, Savicky [21] showed the following.

Lemma 2.30   There are exactly four n-variable symmetric bent functions on $n > 2$ variables. All have degree 2.

Next, we consider a symmetric function, $SB(n, m)$, that was used to analyze the complexity of adders.

Definition 2.31   [18], p. 310

$$SB(n, m) = \sum_{i_1 < i_2 < \ldots < i_m} \oplus \, x_{i_1} x_{i_2} \ldots x_{i_m}, \text{ for } m > 1, \text{ and } SB(n, 0) = 1. \tag{2.3}$$

A special case of $SB(m, n)$ occurs when $m = 2$. In this case, $SB(m, 2)$ is the CQF.

Example 2.32   For $n = 4$, we have

$$
\begin{aligned}
SB(4, 4) &= x_1 x_2 x_3 x_4, \\
SB(4, 3) &= x_1 x_2 x_3 \oplus x_1 x_2 x_4 \oplus x_1 x_3 x_4 \oplus x_2 x_3 x_4, \\
SB(4, 2) &= x_1 x_2 \oplus x_1 x_3 \oplus x_1 x_4 \oplus x_2 x_3 \oplus x_2 x_4 \oplus x_3 x_4, \\
SB(4, 1) &= x_1 \oplus x_2 \oplus x_3 \oplus x_4, \\
SB(4, 0) &= 1.
\end{aligned}
$$

A 4-variable symmetric bent function has the form

$$f = SB(4, 2) \oplus c_1 SB(4, 1) \oplus c_0 SB(4, 0),$$

where $c_0, c_1 \in \{0, 1\}$. Since there are four ways to choose c_1 and c_0, there are four symmetric functions on 4-variables.

However, this suggests a general result. That is, Savicky's [21] result can be stated more precisely, as follows.

Lemma 2.33   There are exactly four n-variable symmetric bent functions on $n > 2$ variables, as follows.

$$f = SB(n, 2) \oplus c_1 SB(n, 1) \oplus c_0 SB(n, 0), \tag{2.4}$$

where $c_0, c_1 \in \{0, 1\}$. Fig. 2.4 shows the distribution of 4-variable symmetric functions according to nonlinearity. There is symmetry about nonlinearity 3. For example, four symmetric functions have nonlinearity 0 ($0, 1, x_1 \oplus x_2 \oplus x_3 \oplus x_4$, and $x_1 \oplus x_2 \oplus x_3 \oplus x_4 \oplus 1$ and four have nonlinearity 6 (and are bent).

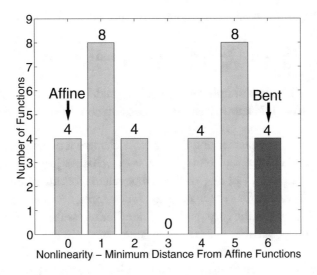

Figure 2.4: Distribution of 4-variable symmetric functions by nonlinearity.

2.4   THE STRICT AVALANCHE CRITERION

Webster and Tavares [25] introduced the following concept.

Definition 2.34   A function f satisfies the **strict avalanche criterion** (**SAC**) iff complementing any single variable complements exactly half of the function values.

Definition 2.35   The **Boolean difference** of a function f with respect to variable x_i is $\frac{df}{dx_i} = f_{x_i=0} \oplus f_{x_i=1}$, where $f_{x_i=\alpha}$ is f with x_i replaced by α ($\in \{0, 1\}$).

Definition 2.36   An n-variable function f is **balanced** iff its weight is 2^{n-1}.

That is, a function is balanced iff its function value has the same number of 1's as 0's. If f satisfies the SAC, then complementing a single variable, x_i complements exactly one half of the function values. It follows that $\frac{df}{dx_i}$ is 1 for all these assignments of values to $x_1, x_2, \ldots,$ and x_n. Further, the converse is true. Thus,

Lemma 2.37   An n-variable function f satisfies the SAC iff $|\frac{df}{dx_i}| = 2^{n-1}$ for all x_i, where $\frac{df}{dx_i}$ is viewed as a function of all n variables.

That is, an n-variable function f satisfies the SAC iff $\frac{df}{dx_i}$ is balanced for all x_i.

Example 2.38   Consider the 4-variable disjoint quadratic function $f = x_1 x_2 \oplus x_3 x_4$. We have

$$\frac{df}{dx_1} = x_2, \frac{df}{dx_2} = x_1, \frac{df}{dx_3} = x_4, \text{ and} \frac{df}{dx_4} = x_3. \tag{2.5}$$

Since each Boolean difference is simply x_i, each is balanced, and the 4-variable disjoint quadratic function satisfies the SAC. It is known that every bent function satisfies the SAC [14].

For some functions, complementing one variable changes a few output values. For example, for the AND function, complementing one variable, say x_1, changes just two output values, those for $x_1 x_2 \ldots x_n = 01 \ldots 1$ and $11 \ldots 1$ or $\frac{1}{2^{n-1}}$ of the output values. For other functions, complementing x_1 changes many output values; for $x_1 \vee x_2 x_3 \ldots x_n$, for example, complementing x_1 changes all but two output values or $1 - \frac{1}{2^{n-1}}$ of the output values. The criterion "avalanche" suggests a small change, such as complementing one variable, yields a much larger change in the output. However, when this is applied to a cryptographic application, the need to achieve maximum confusion suggests that there should be a balance between what is changed and what is not i.e., one-half of the output values are changed). This corresponds to the descriptor "strict". This descriptor is accurate for another reason; the number of functions that satisfy the strict avalanche criterion is small.

Forré [4] introduced the following idea.

Definition 2.39   An n-variable function f satisfies **SAC(k)** iff, for any assignment of values to any k of the n variables, the resulting function satisfies SAC.

Note that $SAC(0)$ is the same as SAC.

Example 2.40   The 4-variable disjoint quadratic function $f = x_1 x_2 \oplus x_3 x_4$ satisfies $SAC(0)$, as shown in Example 2.38, but not $SAC(1)$. For example, if $x_1 = 0$, then complementing x_2 yields no change in the function values, while if $x_1 = 1$, then complementing x_2 changes eight function values.

2.5   THE PROPAGATION CRITERION

A concept closely related to the strong avalanche criterion is the propagation criterion.

Definition 2.41   An n-variable function f satisfies the **propagation criterion ($PC(k)$)** iff complementing any k or fewer of the n variables complements exactly half of the function values.

Note that a function satisfies $PC(1)$ iff it satisfies $SAC(0)$. Indeed, the propagation criterion is a generalization of $SAC(0)$, just as $SAC(k)$ is a generalization of $SAC(0)$.

2.6 CORRELATION IMMUNITY

Another characteristic of Boolean functions that is important in cryptographic applications is correlation immunity. This describes the extent to which the variable values can be guessed given the function value. An example of a function that has low correlation immunity is the AND function on $n > 1$ variables. For example, if this function's output value is 1, then the input variable values are unambiguously $x_1 x_2 \ldots x_n = 11 \ldots 1$. Because of this, the AND function is not a good choice for cryptographic applications.

Interest in correlation immunity developed because Siegenthaler [22] in 1984 showed how an attack can be effectively applied to encryption systems using functions with low correlation immunity.

Definition 2.42 An n-variable function f has **correlation immunity** k iff, for every fixed set S of k variables, $1 \le k \le n$, given the value of f, the probability that S takes on any of its 2^k assignments of values to the k variables is $\frac{1}{2^k}$. If an n-variable function has correlation immunity k and is balanced, then it has **resiliency** of order k.

We expect that the more variable values we know, the greater the chance we know the function value. If we know all values, then we certainly know the function value, since we can examine its truth table. However, we might ask, if we know $n - 1$ of the variable values, do we know the function value? If the function depends on a variable x_i, then there is an assignment of values to the variables besides x_i such that the function changes if x_i changes. Thus, the answer is no. Considering the opposite extreme, we might ask: Does there exist a function such that, for **every** assignment to every set of $n - 1$ variables, we will not be able to determine the function's value? If this is true, then the function has correlation immunity $n - 1$.

An alternative definition of the correlation immunity is as follows.

Definition 2.43 An n-variable function f has **correlation immunity** k iff, for every fixed set S of k variables, $1 \le k \le n$, and for every assignment of values to the variables in S, the weights of all subfunctions are the same.

Now consider several examples.

Definition 2.44 The **barbell function** f_B is $\bar{x}_1 \bar{x}_2 \ldots \bar{x}_n \oplus x_1 x_2 \ldots x_n$.

Definition 2.45 A **threshold function** f_T is 1 iff the weighted sum $\sum_{i=1}^{n} w_i x_i$ exceeds or equals T, where x_i is viewed as an integer equal to its logic value and w_i and T are real numbers.

Example 2.46 It follows that the n-variable AND function has correlation immunity 0 for $n \ge 1$. At the other extreme, the n-variable exclusive-OR function has correlation immunity $n - 1$, for $n \ge 2$. This answers the question posed above. Indeed, there are two functions with correlation immunity $n - 1$, the other being the complement of the exclusive-OR function. There only two functions with correlation immunity greater than that of the two parity functions. These are the constant 0 and 1 functions with correlation immunity n. Note that a function with an odd number of 1's has correlation immunity 0.

The barbell function $\bar{x}_1\bar{x}_2 \ldots \bar{x}_n \oplus x_1 x_2 \ldots x_n$ has correlation immunity 1 for $n \geq 1$.

Any threshold function on $n > 1$ variables has correlation immunity 0 because the probability the function is 1 is different depending on whether or not the value of a variable moves the weighted sum closer to the threshold.

Lemma 2.47   [2] An n-variable function f has correlation immunity 1 iff $f \oplus x_i$ is balanced for all $1 \leq i \leq n$.

Example 2.48   No bent function f has correlation immunity 1 because $f \oplus x_i$ is also bent, and no bent function is balanced.

Lemma 2.49   [2] If the weight of an n-variable function f is not divisible by 2^k, the correlation immunity of f is at most $k - 1$.

For $k = 1$, Lemma 2.49 corresponds to the observation above that a function with an odd number of 1's has correlation immunity 0. If follows that at least half of all functions have correlation immunity 0. For $k = n$, Lemma 2.49 states that a function in which the number of 1's is not divisible by 2^n has correlation immunity at most $n - 1$. There are only two functions in which the number of 1's is divisible by 2^n. These are $f = 0$ and $f = 1$. As observed above, these are the only two functions with correlation immunity n.

2.7   CONCLUDING REMARKS

Bent functions have important cryptographic properties. First, they are very rare. As the number of variables increase, they become a vanishingly small fraction of the total number of functions. Second, there is no formal method of constructing all bent functions. In the research presented here, we have used the sieve technique. In this approach, we generate functions and then test them for bentness. Indeed, we have done this on a reconfigurable computer (SRC Company's SRC-6), in which a large FPGA (a Xilinx Virtex-2, 6000 series) has been configured to enumerate a prospective function, test it against all affine functions generating the distance to each, choose the minimum distance, and tally the generated function according to its nonlinearity.

While general bent functions are difficult to discover, certain specific bent functions can be easily described. For example, the disjoint quadratic function is bent. Further, there are only four bent functions that are totally symmetric and these are easily described.

The number of bent functions is an open question. Preneel [15] showed that the number of 6-variable bent functions is 5,425,430,528 $\sim 2^{32.3}$. For $n = 8$, a very long computation [10] whose results were announced on December 31, 2007 showed that the number of A-classes of bent functions is approximately $2^{97.3}$. Since each A-class has 2^{n+1} functions, there are approximately $2^{106.3}$ bent functions, as shown in Table 2.1.

Although there are no bent functions on 9 variables, there is a surprise regarding the maximum nonlinearity for 9-variable functions. For odd n, one might expect the upper bound on nonlinearity to be described by the "bent concatenation bound" $2^{n-1} - 2^{\frac{n-1}{2}}$, which gives 240 for $n = 9$. In 2006, Kavut, Maitra, and Yücel [7] showed the existence of a 9-variable function with nonlinearity 241. This was improved to nonlinearity 242 in 2008 in Kavut and Yücel [8].

Maitra [11] showed a 13-variable function having nonlinearity 4034, which is 2 greater than the bent concatenation bound, building this function from 16 truth tables of 9-variable bent functions having nonlinearity 242.

Another interesting open question is the largest nonlinearity for n-variable functions, where n is odd.

Still another interesting open question is the largest nonlinearity among balanced functions. This has significance in cryptographic applications because, in practical systems, balance is a dominant requirement. That is, when a bent function is used, it is modified to form a balanced function (which hopefully still has large nonlinearity). The converse problem is to find, among balanced functions, those with maximum nonlinearity. This open question was stated explicitly in Dobbertin and Leander [3]. Unfortunately, the untimely death of the first author stalled publication of this important paper, which is presently available online only [2].

An online database exists that classifies Boolean functions according to nonlinearity, bentness, degree, correlation immunity, propagation criterion, etc. [27].

ACKNOWLEDGMENTS

This research was supported in part by the Grants in Aid for Scientific Research of JSPS, and a grant from the Knowledge Cluster Project of MEXT (Second Stage), and by a contract from the National Security Agency. We are indebted to Pantelimon Stănică of the Naval Postgraduate School, Monterey, CA for comments which led to an improved version of this paper.

2.8   EXERCISES

2.1.   It is known that, for $n \geq 4$, the degree of a bent function of $n = 2r$ variables is at most r [2]. By using this property, show that the number of bent functions of n variables, with n even, is at most

$$2^{\eta(n)},$$

where
$$\eta(n) = 2^{n-1} + \frac{1}{2}\binom{n}{n/2}.$$

2.2. By using the result of Problem 2.1, calculate an upper bound on the number of n variable bent functions, where $n = 4, 6$, and 8.

2.3. Show that the number of n-variable homogeneous functions with degree k is
$$2^{\binom{n}{k}} - 1.$$

2.4. By using the result of Problem 2.3, show that the number of n-variable homogenous functions of any degree is
$$\sum_{k=0}^{n} 2^{\binom{n}{k}} - n.$$

2.5. By using the result of Problem 2.4, calculate the number of n-variable homogenous functions of any degree for $n = 4, 6$, and 8.

2.6. Show that the number of n-variable balanced functions is
$$\binom{2^n}{2^{n-1}}.$$

2.7. By using the result of Problem 2.6, calculate the number of n-variable balanced functions for $n = 4$ and 6.

2.8. Explain why balanced functions are preferred for the circuits in cryptography.

2.9. Show that there are more bent functions than symmetric functions and affine functions.

2.10. Let s_ϕ be the first coefficient of the Walsh spectrum. Let $w(f)$ be the weight of the function f, i.e., the number of combinations that are mapped to 1. Then, show the relation.
$$s_\phi = w(\bar{f}) - w(f) = 2^n - 2w(f).$$

2.11. Let s_ϕ be the first coefficient of the Walsh spectrum. Show that, for balanced functions, $s_\phi = 0$.

2.12. Prove that a bent function has a flat Walsh spectrum.

2.13. Let $w(f)$ be the weight of f. Prove $w(DQF) = 2^{n-1} - 2^{\frac{n}{2}-1}$.

2.14. *Prove:* $DQF(x) = x_1x_2 \oplus x_3x_4 \oplus \ldots \oplus x_{n-1}x_n$ *is bent.*

2.15. Let the weight of a function f be $w(f)$. Then, show that the non-linearity of f is at most $w(f)$.

2.16. *Prove: If $f(x)$ is bent and $a(x)$ is affine, then $f(x) \oplus a(x)$ is bent.*

2.17. *Prove: If $f(x)$ is bent, then so is $f(x)|_{\bar{x}_i \to x_i}$.*

REFERENCES

[1] C. Carlet and A. Klapper, "Upper bounds on the numbers of resilient functions and of bent functions," *Proc. of the 23rd Symposium on Information Theory in the Benelux*, Louvain-La-Neuve, Belgique, Mai 2002 Publie par "Werkgemeeschal voor Informatie-en Communicatietheorie, Enschede, The Nederlands, pp. 307-314, 2002. 2.2

[2] T. W. Cusick and P. Stănică, *Cryptographic Boolean Functions and Applications*, Academic Press, San Diego, CA, 2009. 1, 2.2, 2.20, 2.21, 2.47, 2.49, 2.7, 2.1

[3] H. Dobbertin and G. Leander, "Crytographer's toolkit for construction of 8-bit bent functions," preprint: http://eprint.iacr.org/2005/089.pdf 2.7

[4] R. Forré, "The strict avalanche criterion: Spectral properties of Boolean functions and an extended definition," *Advances in cryptology, Crypto'88*, pp. 450-468. 2.4

[5] K. J. Horadam, *Hadamard Matrices and Their Applications*, Princeton University Press, 2007.

[6] M. G. Karpovsky, R. S. Stankovic, and J. T. Astola, *Spectral Logic and Its Applications for the Design of Digital Devices*, Wiley-Interscience, 2008. DOI: 10.1002/9780470289228

[7] S. Kavut, S. Maitra, and M. D. Yücel, "There exist Boolean functions on n (odd) variables having nonlinearity $> 2^{n-1} - 2^{\frac{n-1}{2}}$ iff $n > 7$," *IEEE Trans. on Infor. Theory*, Vol. 53, No. 5, pp. 1743-1751, May 2007. 2.7

[8] S. Kavut and M. D. Yücel, "9-variable Boolean functions with nonlinearity 242 in the generalized rotation class," *Cryptology EPrint* Report 2006/181, 28 May, 2006. http://eprint.iacr.org/2006/131 2.7

[9] D. Knuth, *The Art of Computer Programming*, Vol. 4, Fascicle 0, "Introduction to combinatorial algorithms and Boolean functions," Addison-Wesley Publishing Company, pp. 95-96 and p. 180, 2008. 2.2

[10] P. Langevin, G. Leander, P. Rabizzoni, P. Véron, J.-P. Zanotti, "Classification of Boolean quartics forms in eight variables," http://langevin.univ-tln.fr/project/quartics/, Dec. 2007. 2.2, 2.7

[11] S. Maitra, "Balanced Boolean functions on 13 variables having nonlinearity strictly greater than the bent concatenation bound," *Cryptology EPrint*, Report 2007/309, 2007, http://eprint.iacr.org/2007/309.pdf 2.7

[12] M. Matsui, "The first experimental cryptanalysis of the Data Encryption Standard," *Advances in Cryptology CRYPTO '94* (*Lecture Notes in Computer Sciences*, no. 839), Springer-Verlag, pp. 1-11, 1994. 2.2

[13] M. Matsui, "Linear cryptanalysis method for DES cipher," *Advances in Cryptology EURO-CRYPT '93* (*Lecture Notes in Computer Sciences*, no. 765), Springer-Verlag, pp. 386-397, 1994. 2.2

[14] W. Meier and O. Staffelbach, "Nonlinearity criteria for cryptographic functions," *Advances in Cryptology EUROCRYPT '89* (*Lecture Notes in Computer Sciences*, no. 434), Springer-Verlag, pp. 549-562, 1990. 2.2, 2.9, 2.38

[15] B. Preneel, *Analysis and Design of Cryptographic Hash Functions*, Ph.D. Thesis, Katholieke Universiteit Leuven, K. Mercierlaan 94, 3001 Leuven, Belguim, 1993. 2.2, 2.7

[16] C. Qu, J. Seberry, and J. Pieprzyk, "Homogeneous bent functions," *Discrete Applied Math.*, Vol. 102, pp. 133-139, 2000. DOI: 10.1016/S0166-218X(99)00234-6 2.2, 2.2

[17] O. S. Rothaus, "On 'bent' functions," *Journal of Combinatorial Theory*, Ser. A, 20, pp. 300-305, Nov. 1976. 2.2, 2.2, 2.2, 2.2, 2.2, 2.2, 2.2, 2.2

[18] T. Sasao (ed.), *Logic Synthesis and Optimization*, Kluwer Academic Publishers, 1993. 2.31

[19] T. Sasao and M. Fujita (ed.), *Representation of Discrete Functions*, Kluwer Academic Publishers, 1996. 2.2

[20] T. Sasao, *Switching Theory for Logic Synthesis*, Kluwer Academic Publishers, 1999. 2.2

[21] P. Savicky, "On bent functions that are symmetric," *European J. of Combinatorials*, 15, pp. 407-410, 1994. 2.3, 2.3

[22] T. Siegenthaler, "Correlation immunity of nonlinear combining functions for cryptographic applications," *IEEE Trans. on Information Theory*, IT-30(5), pp. 776-780, Sept. 1984. DOI: 10.1109/TIT.1984.1056949 2.6

[23] X. Wang, J. Zhou, and Y. Zang, "A note on homogeneous bent functions," *Eighth Inter. Conf. on Software Eng., Artificial Intelligence, Networking, and Parallel/Distributed Computing*, pp. 138-142, 2007. 2.2

[24] I. Wegener, *Branching Programs and Binary Decision Diagrams: Theory and Applications*, SIAM Monograph on Discrete Mathematics and Applications, Philadelphia, 2000. 2.23

[25] I. Webster and S. E. Tavares, "On the design of S-boxes," *Advances in Cryptology - Crypto '85 (1986)*, Vol. 218, Lecture Notes in Computer Science, Springer, Berlin, pp. 523-534, 1987. 2.4

[26] T. Xia, J. Seberry, J. Pieprzyk, and C. Charnes, "Homogeneous bent functions of degree n in $2n$ variables do not exist for $n > 3$," *Discrete Applied Math.*, Vol. 142, pp. 127-132, 2004. DOI: 10.1016/j.dam.2004.02.006 2.2, 2.2, 2.2

[27] *Universitetet i Bergen, Institutt for informatikk*, Bergen, Norway. Online database of Boolean functions according to bentness, degree, correlation immunity, propagation criterion, etc. `http://www.ii.uib.no/~$mohamedaa/odbf/search.html` 2.7

CHAPTER 3

Boolean Differential Calculus

Bernd Steinbach and Christian Posthoff

CHAPTER SUMMARY

Boolean Differential Calculus (BDC) extends Boolean algebra. While Boolean algebra is focused on values of logic functions, BDC allows the evaluation of *changes* of the function values. Such changes can be investigated between certain pairs of function values as well as regarding whole subspaces. Due to the same basic data structures, BDC can be applied to any task described by logic functions and equations together with the Boolean algebra. Widely used is BDC in analysis, synthesis, and testing of digital circuits. In this chapter, we introduce basic definitions of BDC together with some typical applications.

3.1 INTRODUCTION

Practical problems inspired the development of the Boolean Differential Calculus (BDC). The increasing size of digital circuits in the 1950's required approaches to detect faults in digital devices. Original ideas of *Reed* [8], *Huffman* [5] and *Akers* [1] dealt with this special problem. *Thayse* and his colleagues presented the basic theoretical concepts in comprehensive publications, such as [9] and [18]. A research center dealing with these theories and applications was the Chemnitz University of Technology (Germany). In 1981, a monograph [3] was published. A comprehensive presentation of BDC was written in 2004 as a part of [11] and complemented in 2009 by the book [16], which contains many examples and exercises together with their solutions.

Why do we need an extension of Boolean Algebra by BDC? The definitions and laws of the traditional Boolean Algebra apply to static *values* of logic functions. However, it is also desirable to use other logic functions to describe *changes* of the values of the given logic functions in a simple way. Traditional Boolean Algebra does not express such changes explicitly, but only in a complicated implicit manner. BDC avoids these restrictions and opens the view to a new, more appropriate and more natural terminology which is suitable to describe problems on a higher level of abstraction [15].

The applications of BDC include:

- *Fault detection*: In order to detect faults in a circuit, the difference between an output value of the fault-free circuit and the circuit with a fault is evaluated [1], [4], [5], [8], [9], [11], [17].

- *Detection of redundant variables*: A logic function may be specified by a logic expression. A logic variable can be removed from such an expression if the change of the value of this variable does not change the function values.

- *Detection of hazards*: During the transition between two input patterns with the same output signal, an intermediate inverse signal value may appear on the output of the combinational part that can perturb the required behavior of an asynchronous sequential circuit [2], [3], [11], [16]. How can such possible intermediate signal changes (called hazards) be detected?

- *Application to logic synthesis*: The description of the behavior of a circuit can be incomplete. It will save time when a simple analysis finds out whether at least one circuit structure exists for a given behavior description. In such investigations, the differences between allowed and forbidden behavior must be considered, and the realizability of the behavior must hold for all inputs.

- *Functional decomposition*: The minimization of a multi-level circuit is a complex task. By using the BDC, this task has been solved at least in such a way that subfunctions that do not depend on several variables of the composed function, can be detected efficiently [6], [7], [10], [11], [13], [14], [16], [17], [20].

- *Solution of Boolean equations with regard to variables* $\mathbf{y}$: Are there uniquely specified output functions $y_i = f_i(\mathbf{x})$ for a circuit behavior given by the system equation $F(\mathbf{x}, \mathbf{y}) = \mathbf{1}$? Solutions are given in [2], [3], [11], [16], [19].

A comprehensive software package XBOOLE [12], [16] with several versions has been developed which implemented BDC operations. It restricts the efforts of researchers and applicants to the modeling of the problem while the software "is doing the rest of the work."

3.2   PRELIMINARIES

In the following, we will use the Shannon decomposition of a logic function $f(x_i, \mathbf{x_1})$ with regard to x_i which is given by

$$f(x_i, \mathbf{x_1}) = \overline{x}_i \wedge f(x_i = 0, \mathbf{x_1}) \vee x_i \wedge f(x_i = 1, \mathbf{x_1}), \tag{3.1}$$

where $f(x_i = 0, \mathbf{x_1})$ is called the negative cofactor and $f(x_i = 1, \mathbf{x_1})$ the positive cofactor. The variable x_i and the associated cofactor are connected by the AND-operation which is emphasized by the sign '$\wedge$'. The sign of the AND-operation will often be left out (in the same way as the multiplication sign in other fields of mathematics). $\mathbf{x_1}$ summarizes the remaining variables in a certain order into a vector. Assume for example $\mathbf{x_1} = (x_{11}, x_{12}, x_{13})$, then $f(1011) = 0$ means that the function value of f is equal to zero for the values of the variables $x_i = 1, x_{11} = 0, x_{12} = 1$, and $x_{13} = 1$.

Let be $f(\mathbf{x})$ and $g(\mathbf{x})$ logic functions. Then, the relation $f(\mathbf{x}) \leq g(\mathbf{x})$ is equivalent to the equation $f(\mathbf{x}) \wedge \overline{g(\mathbf{x})} = 0$:

$$f(\mathbf{x}) \leq g(\mathbf{x}) \quad \Leftrightarrow \quad f(\mathbf{x}) \wedge \overline{g(\mathbf{x})} = 0. \tag{3.2}$$

The relation $f(\mathbf{x}) \leq g(\mathbf{x})$ holds if all assignments $\mathbf{a}$ of values to $\mathbf{x}$ have the property $(f(\mathbf{a}), g(\mathbf{a})) \in \{(0,0), (0,1), (1,1)\}$. These three pairs also satisfy the given equation. The remaining pair $(1,0)$ is not a solution of the equation.

Similarly, the equation $f(\mathbf{x}) = g(\mathbf{x})$ holds for the pairs of function values $(f(\mathbf{a}), g(\mathbf{a})) \in \{(0,0), (1,1)\}$. Therefore, we can use the relation $f(\mathbf{x}) \leq g(\mathbf{x})$ together with the equation $\overline{f(\mathbf{x})} \wedge g(\mathbf{x}) = 0$, which excludes the pair $(0,1)$:

$$f(\mathbf{x}) = g(\mathbf{x}) \quad \Leftrightarrow \quad f(\mathbf{x}) \leq g(\mathbf{x}) \text{ and } \overline{f(\mathbf{x})} \wedge g(\mathbf{x}) = 0. \tag{3.3}$$

3.3 SIMPLE DERIVATIVE OPERATIONS

We start with the simplest operations of BDC. We change the value of one variable $x_i \in \{x_1, \ldots, x_n\}$ (from 0 to 1 or vice versa) and observe what happens to the value of a given function depending on this variable. This can be done by combining the positive and the negative cofactor of the given function by means of EXOR ($\oplus$), AND ($\wedge$) or OR ($\vee$).

Definition 3.1 Let $f(\mathbf{x}) = f(x_i, \mathbf{x}_1)$ be a logic function of n variables. Then

$$\frac{\partial f(\mathbf{x})}{\partial x_i} = f(x_i = 0, \mathbf{x}_1) \oplus f(x_i = 1, \mathbf{x}_1) \tag{3.4}$$

is the (simple) derivative,

$$\min_{x_i} f(\mathbf{x}) = f(x_i = 0, \mathbf{x}_1) \wedge f(x_i = 1, \mathbf{x}_1) \tag{3.5}$$

the (simple) minimum and

$$\max_{x_i} f(\mathbf{x}) = f(x_i = 0, \mathbf{x}_1) \vee f(x_i = 1, \mathbf{x}_1) \tag{3.6}$$

the (simple) maximum of the logic function $f(\mathbf{x})$ with regard to the variable x_i.

The result of any derivative operations is a new logic function g which reflects certain properties of a given function f. In the case of a simple derivative operation, the result does not depend anymore on the variable x_i (x_i has been set to 0 and to 1).

The *derivative* is equal to 1 iff the change of the variable x_i changes the function value for constant values of the remaining variables $\mathbf{x}_1$ (since $0 \oplus 1 = 1 \oplus 0 = 1$); otherwise, it is equal to 0.

The *minimum* is equal to 1 iff the change of the variable x_i does not change the function value 1 for constant values of $\mathbf{x_1}$ (since $1 \wedge 1 = 1$); otherwise, it is equal to 0. We also can say that the minimum is equal to 1 for such subspaces $\mathbf{x_1} = const$ where all function values are equal to 1.

The *maximum* is equal to 0 iff the change of the variable x_i does not change the function value 0 for constant values of $\mathbf{x_1}$ (since $0 \vee 0 = 0$); otherwise, it is equal to 1. From another point of view, the maximum is equal to 1 for such subspaces $\mathbf{x_1} = const$ where at least one function value is equal to 1.

Figure 3.1 shows an example of a function $f(x_1, x_2, x_3)$ and the derivative operations with regard to x_3 depicted by Karnaugh maps. The arrows in the Karnaugh map of $f(x_1, x_2, x_3)$ indicate the values that must be combined by $\wedge, \oplus$ or $\vee$. Note that the results of all simple derivative operations do not depend on x_3.

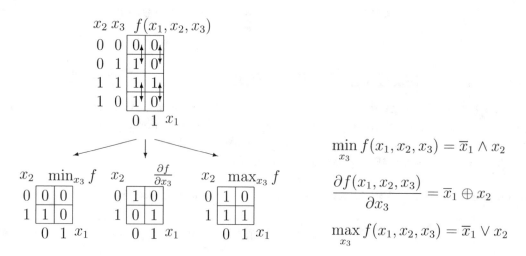

Figure 3.1: The Karnaugh maps of $f(x_1, x_2, x_3) = \overline{x}_1 (x_2 \vee x_3) \vee x_2 x_3$ and all simple derivative operations with regard to $x_i = x_3$.

By using the three derivative operations, it can be calculated whether the function values are constant equal to 0, equal to 1 or change from 0 to 1 in a given subspace where only changes of x_i are considered (x_3 in the example).

It is sometimes convenient to use another form of the definitions instead of (3.4), (3.5), and (3.6):

$$\frac{\partial f(\mathbf{x})}{\partial x_i} = f(x_i, \mathbf{x_1}) \oplus f(\overline{x}_i, \mathbf{x_1}), \tag{3.7}$$

$$\min_{x_i} f(\mathbf{x}) = f(x_i, \mathbf{x_1}) \wedge f(\overline{x}_i, \mathbf{x_1}), \tag{3.8}$$

$$\max_{x_i} f(\mathbf{x}) = f(x_i, \mathbf{x_1}) \vee f(\overline{x}_i, \mathbf{x_1}). \tag{3.9}$$

The definitions in (3.4), (3.5), and (3.6) emphasize the explicit use of the constants 0 and 1 for x_i, in any expression for f. In (3.7), (3.8), and (3.9), the use of x_i and $\overline{x}_i$ also implements the transition from 0 to 1. In this case, the variable x_i will be replaced in any expression for f by $\overline{x}_i$ and vice versa. Then, the two expressions also will be combined and simplified.

The terms *minimum* and *maximum* of the two other derivative operations are motivated by the following theorem:

Theorem 3.2

$$\min_{x_i} f(x_i, \mathbf{x}) \leq f(x_i, \mathbf{x}) \leq \max_{x_i} f(x_i, \mathbf{x}). \tag{3.10}$$

Proof 3.3   We prove the left side of this theorem. In order to make the comparison meaningful, $\min_{x_i} f(\mathbf{x})$ must be understood as a function of the variables $x_1, \ldots, x_n$ including x_i. This is always possible because of $(x_i \vee \overline{x}_i) \min_{x_i} f(\mathbf{x}) = \min_{x_i} f(\mathbf{x})$. We use $f(x_i = 0, \mathbf{x}_1) = f(0)$ and $f(x_i = 1, \mathbf{x}_1) = f(1)$ for the two cofactors and replace the $\leq$-relation by the equivalent equation (see 3.2):

$$\min_{x_i} f(\mathbf{x}) \;\; \leq \;\; f(\mathbf{x}) \;\; \Leftrightarrow \;\; \min_{x_i} f(\mathbf{x}) \wedge \overline{f(\mathbf{x})} = 0. \tag{3.11}$$

Then, we replace in (3.11) $\min_{x_i} f(\mathbf{x})$ by its definition (3.5) and $f(\mathbf{x})$ by its Shannon decomposition with regard to x_i (3.1) and simplify the expression step by step:

$$
\begin{aligned}
\min_{x_i} & f(\mathbf{x}) \wedge \overline{f(\mathbf{x})} \\
&= \; f(0)f(1) \wedge \overline{\overline{x}_i f(0) \vee x_i f(1)}, \\
&= \; f(0)f(1) \wedge \overline{\overline{x}_i f(0)} \wedge \overline{x_i f(1)}, \\
&= \; f(0)f(1) \wedge (x_i \vee \overline{f(0)}) \wedge (\overline{x}_i \vee \overline{f(1)}), \\
&= \; f(0)f(1) \wedge (x_i \, \overline{x}_i \vee x_i \, \overline{f(1)} \vee \overline{x}_i \, \overline{f(0)} \vee \overline{f(0)} \; \overline{f(1)}), \\
&= \; f(0)f(1) \wedge x_i \, \overline{x}_i = 0.
\end{aligned}
\tag{3.12}
$$

The right inequality of Theorem 3.2 should be proven as Exercise 3.1, in a similar way.

The simple derivative operations can be generalized in two different ways. Both of them are needed in practical applications. On the one hand, we can use a sequence of simple derivative operations with regard to single variables. These so-called m-fold derivative operations can be defined directly using the definition of simple derivative operations as shown in Section 3.5.

On the other hand, the simple derivative operations compare two function values that are reached by changing a single variable. The change can involve more than one variable at the same time. This leads to vectorial derivative operations, as shown in Section 3.4.

3.4   VECTORIAL DERIVATIVE OPERATIONS

The theory can accommodate a change in more than one variable. One generalization is that all variables of the vector $\mathbf{x}_0$ can be changed at the same time. The vectorial derivative operations describe the relationship between the associated pairs of function values.

Definition 3.4   Let $f(\mathbf{x}) = f(\mathbf{x}_0, \mathbf{x}_1)$ be a logic function of n variables. Then

$$\frac{\partial f(\mathbf{x}_0, \mathbf{x}_1)}{\partial \mathbf{x}_0} = f(\mathbf{x}_0, \mathbf{x}_1) \oplus f(\overline{\mathbf{x}}_0, \mathbf{x}_1) \tag{3.13}$$

is the vectorial derivative,

$$\min_{\mathbf{x}_0} f(\mathbf{x}_0, \mathbf{x}_1) = f(\mathbf{x}_0, \mathbf{x}_1) \wedge f(\overline{\mathbf{x}}_0, \mathbf{x}_1) \tag{3.14}$$

the vectorial minimum, and

$$\max_{\mathbf{x}_0} f(\mathbf{x}_0, \mathbf{x}_1) = f(\mathbf{x}_0, \mathbf{x}_1) \vee f(\overline{\mathbf{x}}_0, \mathbf{x}_1) \tag{3.15}$$

the vectorial maximum of the logic function $f(\mathbf{x}_0, \mathbf{x}_1)$ with regard to the variables of $\mathbf{x}_0$.

From their definition, all vectorial derivative operations result in logic functions that depend in general on both the variables of $\mathbf{x}_0$, which have been included in the derivative operation and the variables of $\mathbf{x}_1$. There are certain functions $f(\mathbf{x}_0, \mathbf{x}_1)$ where the result of a vectorial derivative operation does not depend on some of these variables.

The vectorial derivative is equal to 1 iff the simultaneous change of the variables of $\mathbf{x}_0$ changes the function value for fixed values of the remaining variables of $\mathbf{x}_1$. This is caused by the EXOR-operation in (3.13) that is equal to 1 if different pairs of function values occur, e.g., (01) or (10) occur.

The vectorial minimum is equal to 1 iff the simultaneous change of the variables of $\mathbf{x}_0$ does not change the function value 1 for fixed values of the remaining variables of $\mathbf{x}_1$. This is caused by the AND-operation in (3.14) that is equal to 1 only if the pair of function values (11) occurs.

The vectorial maximum is equal to 1 iff the simultaneous change of the variables $\mathbf{x}_0$ yields at least one function value 1. On the other hand, the vectorial maximum is equal to 0 if the common change of the variables of $\mathbf{x}_0$ does not change the function value 0 for fixed values of the remaining variables of $\mathbf{x}_1$. This is caused by the OR-operation in (3.15).

Figure 3.2 shows an example of the function $f(x_1, x_2, x_3)$ and all vectorial derivative operations with regard to $\mathbf{x}_0 = (x_2, x_3)$, as depicted by Karnaugh maps. The arrows in the Karnaugh map of $f(x_1, x_2, x_3)$ indicate the values that must be combined using an AND-, an EXOR-, or an OR-operation to get the vectorial minimum, the vectorial derivative or the vectorial maximum of $f(x_1, x_2, x_3)$ with regard to the change direction (x_2, x_3). Figure 3.2 shows that both the vectorial minimum and the vectorial maximum depend on all variables. In general, also the vectorial derivative depends on all variables.

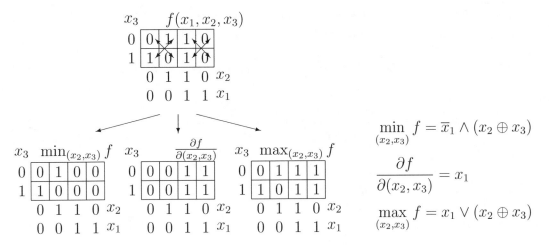

Figure 3.2: The Karnaugh maps of the function $f(x_1, x_2, x_3) = \overline{x}_1\,\overline{x}_2\,x_3 \vee x_1\,x_2 \vee x_2\,\overline{x}_3$ and all vectorial derivatives with regard to $\mathbf{x}_0 = (x_2, x_3)$.

In the case of vectorial derivative operations, the pairs of function values to be evaluated are obtained by changing the values of all variables of $\mathbf{x}_0$ simultaneously. In this regard, the simple derivative operations are special cases of the vectorial derivative operations where the vector $\mathbf{x_0}$ contains only one variable. Similar theorems hold for both simple and vectorial derivative operations.

Theorem 3.5

$$\min_{\mathbf{x}_0} f(\mathbf{x}_0, \mathbf{x}_1) \le f(\mathbf{x}_0, \mathbf{x}_1) \le \max_{\mathbf{x}_0} f(\mathbf{x}_0, \mathbf{x}_1). \tag{3.16}$$

The proof of both inequalities of (3.16) should be done as Exercise 3.3 in a similar manner as shown above for Theorem 3.2.

3.5   *m*-FOLD DERIVATIVE OPERATIONS

The result of any simple derivative operation of a logic function with regard to a single variable x_i is again a logic function that does not depend on x_i (it is now a function of $n - 1$ variables). Hence, from a formal point of view, it is possible to use the result of a simple derivative operation with regard to x_i and apply the same derivative operation with regard to another variable x_j. By repeating this approach m times, we get m-fold derivative operations, which are very helpful in many applications, because the results have a smaller number (up to m) of variables [11], [16].

Definition 3.6   Let $f(\mathbf{x}) = f(\mathbf{x}_0, \mathbf{x}_1)$ be a logic function of n variables, and let $\mathbf{x}_0 = (x_1, x_2, \ldots, x_m)$. Then

$$\frac{\partial^m f(\mathbf{x}_0, \mathbf{x}_1)}{\partial x_1 \partial x_2 \ldots \partial x_m} = \frac{\partial}{\partial x_m}(\ldots(\frac{\partial}{\partial x_2}(\frac{\partial f(\mathbf{x}_0, \mathbf{x}_1)}{\partial x_1}))\ldots) \tag{3.17}$$

is the m-fold derivative,

$$\min_{\mathbf{x}_0}^m f(\mathbf{x}_0, \mathbf{x}_1) = \min_{x_m}(\ldots(\min_{x_2}(\min_{x_1} f(\mathbf{x}_0, \mathbf{x}_1)))\ldots) \tag{3.18}$$

the m-fold minimum,

$$\max_{\mathbf{x}_0}^m f(\mathbf{x}_0, \mathbf{x}_1) = \max_{x_m}(\ldots(\max_{x_2}(\max_{x_1} f(\mathbf{x}_0, \mathbf{x}_1)))\ldots) \tag{3.19}$$

the m-fold maximum and

$$\Delta_{\mathbf{x}_0} f(\mathbf{x}_0, \mathbf{x}_1) = \min_{\mathbf{x}_0}^m f(\mathbf{x}_0, \mathbf{x}_1) \oplus \max_{\mathbf{x}_0}^m f(\mathbf{x}_0, \mathbf{x}_1) \tag{3.20}$$

the Δ - operation of the function $f(\mathbf{x}_0, \mathbf{x}_1)$ with regard to the set of variables $\mathbf{x}_0$.

In contrast to the vectorial derivative operations where pairs of function values decide the result, the results are now determined by all 2^m function values of a subspace $\mathbf{x}_1 = const$ in case of all m-fold derivative operations. Due to Definitions 3.1 and 3.6, the results of all m-fold derivative operations with regard to $\mathbf{x}_0$ do not depend on the variables of $\mathbf{x}_0$. The variables of the vector $\mathbf{x}_0$ can be used in any order. The result of any m-fold derivative operation will always be the same, since the respective operations are commutative [3].

The m-fold derivative of a function $f(\mathbf{x}_0, \mathbf{x}_1)$ with regard to $\mathbf{x}_0$ is equal to 1 for such subspaces $\mathbf{x}_1 = const$ that include an odd number of function values 1.

The m-fold minimum of a function $f(\mathbf{x}_0, \mathbf{x}_1)$ with regard to $\mathbf{x}_0$ is equal to 1 for such subspaces $\mathbf{x}_1 = const$ that include for all $\mathbf{x}_0$ function values 1.

The m-fold maximum of a function $f(\mathbf{x}_0, \mathbf{x}_1)$ with regard to $\mathbf{x}_0$ is equal to 1 for such subspaces $\mathbf{x}_1 = const$ that include at least one function value 1.

Due to Definitions (3.18), (3.19), and (3.20), the Δ - operation of a function $f(\mathbf{x}_0, \mathbf{x}_1)$ with regard to $\mathbf{x}_0$ is equal to 1 for such subspaces $\mathbf{x}_1 = const$ that include both function values 0 and 1. Conversely, the Δ - operation is equal to 0 if the subspaces $\mathbf{x}_1 = const$ include the same function value for all $\mathbf{x}_0$.

Figure 3.3 shows, as an example, the Karnaugh maps of the logic function $f(x_1, x_2, x_3, x_4)$ and all possible m-fold derivative operations, for $m = 2$ and $\mathbf{x}_0 = (x_2, x_4)$. The arrows in the Karnaugh map of $f(x_1, x_2, x_3, x_4)$ and the simple derivative operations with regard to x_4 indicate the values

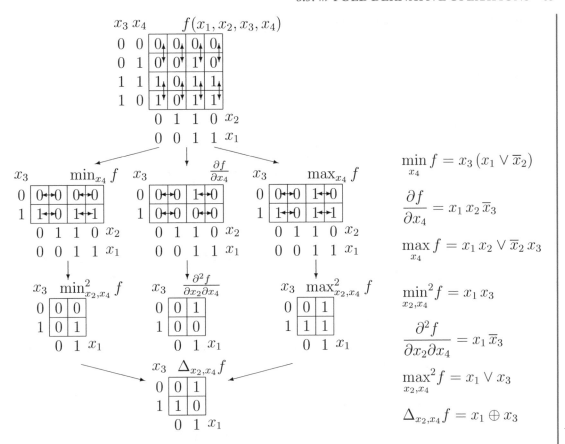

Figure 3.3: The Karnaugh maps of the function $f(x_1, x_2, x_3, x_4) = x_3(x_1 \vee \overline{x}_2) \vee x_1 x_2 \overline{x}_3 x_4$ and all 2-fold derivatives with regard to (x_2, x_4).

that must be combined using an AND-, an EXOR-, or an OR-operation in order to get finally the *m*-fold minimum, the *m*-fold derivative, the *m*-fold maximum or the Δ - operation of $f(x_1, x_2, x_3, x_4)$ with regard to (x_2, x_4). Figure 3.3 shows that the results of all *m*-fold derivative operations do not depend on (x_2, x_4).

The comparison between the Karnaugh maps in Figure 3.3 of the 2-fold derivative and the Δ - operation of the same function with regard to the same variables emphasizes the different evaluation of the values in the subspaces $(x_1, x_3) = const$ of these *m*-fold derivative operations.

A further property of the *m*-fold minimum and the *m*-fold maximum can be observed in Figure 3.3. It is the so-called projection property and will be utilized in many applications. Any value 0 in a subspace $\mathbf{x}_1 = const$ is projected to the result value 0 for the associated subspace of the

m-fold minimum. In case of an m-fold maximum, any value 1 in a subspace $\mathbf{x}_1 = const$ is projected to the result value 1 for the associated subspace.

The more variables belong to the set of variables $\mathbf{x}_0$, the smaller is the m-fold minimum and the larger the m-fold maximum.

Theorem 3.7

$$\min_{\mathbf{x}_0}^m f(\mathbf{x}_0, \mathbf{x}_1) \leq \min_{x_i} f(\mathbf{x}_0, \mathbf{x}_1) \leq f(\mathbf{x}_0, \mathbf{x}_1),$$
$$f(\mathbf{x}_0, \mathbf{x}_1) \leq \max_{x_i} f(\mathbf{x}_0, \mathbf{x}_1) \leq \max_{\mathbf{x}_0}^m f(\mathbf{x}_0, \mathbf{x}_1). \tag{3.21}$$

Proof 3.8   Both inequalities in the middle of (3.21) are valid due to Theorem 3.2. This is the proof of the leftmost inequality of (3.21), which can be expressed as a sequence of inequalities as follows:

$$\min_{\mathbf{x}_0}^m f(\mathbf{x}) \leq \min_{\mathbf{x}_0 \backslash x_{0m}}^{m-1} f(\mathbf{x}) \leq \dots \leq \min_{(x_{02}, x_i)}^2 f(\mathbf{x}) \leq \min_{x_i} f(\mathbf{x}). \tag{3.22}$$

Without loss of generality each inequality of (3.22), and consequently the leftmost inequality of (3.21) holds if the following holds:

$$\min_{\mathbf{x}_{0k}, x_{0ik}}^{k+1} f(x_{0ik}, \mathbf{x}_{0k}, \mathbf{x}_1) \leq \min_{\mathbf{x}_{0k}}^k f(x_{0ik}, \mathbf{x}_{0k}, \mathbf{x}_1). \tag{3.23}$$

We substitute $\min_{\mathbf{x}_{0k}}^k f(x_{0ik}, \mathbf{x}_{0k}, \mathbf{x}_1) = g(x_{0ik}, \mathbf{x}_1)$ and get due to (3.18):

$$\begin{aligned}
\min_{\mathbf{x}_{0k}, x_{0ik}}^{k+1} f(x_{0ik}, \mathbf{x}_{0k}, \mathbf{x}_1) &\leq \min_{\mathbf{x}_{0k}}^k f(x_{0ik}, \mathbf{x}_{0k}, \mathbf{x}_1), \\
\min_{x_{0ik}} \left[\min_{\mathbf{x}_{0k}}^k f(x_{0ik}, \mathbf{x}_{0k}, \mathbf{x}_1) \right] &\leq g(x_{0ik}, \mathbf{x}_1), \\
\min_{x_{0ik}} g(x_{0ik}, \mathbf{x}_1) &\leq g(x_{0ik}, \mathbf{x}_1).
\end{aligned} \tag{3.24}$$

Due to Theorem 3.2, (3.24) proves the leftmost inequality of (3.21).

The proof of the rightmost inequality of Theorem 3.7 should be done as Exercise 3.4, in a similar manner as shown above for the leftmost inequality of the same theorem.

3.6   APPLICATIONS TO HAZARD DETECTION

A glitch is an actual occurrence of a spurious signal of a circuit. A hazard is the possibility that a glitch can occur. BDC allows us to express the conditions of any hazard on a high level of abstraction.

Consider an asynchronous sequential circuit that counts how often the circuit output of a function $f(\mathbf{x})$ changes its value from 0 to 1. Each glitch perturbs the required behavior of this sequential circuit.

A glitch causes an intermediate inverse signal value. Depending on the normal behavior and the assumption of a glitch, we consider four types of hazards,

- (010) - static 0 - hazard,

- (101) - static 1 - hazard,

- (0101) - dynamic 0 - hazard, and

- (1010) - dynamic 1 - hazard,

where each binary vector represents a time sequence of logic values. Furthermore, we distinguish between *functional hazards* and *structural hazards*. Functional hazards denote the conditions of hazards based on the function without any dependence of a associated circuit structure. Structural hazards depend on the realized circuit. Hence, different structural hazards can exist for different circuits of the same function.

First, we model the conditions of **functional hazards** on a high level using the derivative operations of the Boolean Differential Calculus. A sequence of three function values of a static hazard requires the change of the values of at least two variables. A value change of the variables x_i, x_j at the same time can cause a static hazard if the function satisfies the following conditions:

1. The function has the same value before and after the change. That means the function has the same value even though the values of x_i and x_j change at the same time.

2. Different function values exist in the respective subspace such that the function can switch to the opposite value for a short period of time.

Hence, all patterns $(x_i, x_j, \mathbf{x}_1)$ for which such static hazards may appear are the solution of the following equation:

$$\overline{\frac{\partial f(x_i, x_j, \mathbf{x}_1)}{\partial (x_i, x_j)}} \wedge \Delta_{(x_i, x_j)} f(x_i, x_j, \mathbf{x}_1) = 1. \tag{3.25}$$

The complement of the vectorial derivative of $f(x_i, x_j, \mathbf{x}_1)$ with regard to (x_i, x_j) is 1 iff the first condition holds, and the Δ - operation in (3.25) with regard to (x_i, x_j) is 1 iff the second condition holds.

Example 3.9   Consider the function $f(x_1, x_2, x_3, x_4)$ of Figure 3.3. Suppose that (x_2, x_4) change at the same time and static hazards must be found:

$$\overline{\frac{\partial f(x_1, x_2, x_3, x_4)}{\partial (x_2, x_4)}} = x_1 x_3 \vee \overline{x}_1 \overline{x}_3 \vee x_1 \overline{x}_3 (x_2 \oplus x_4), \tag{3.26}$$

$$\Delta_{x_2, x_4} f(x_1, x_2, x_3, x_4) = x_1 \oplus x_3, \tag{3.27}$$

$$\overline{\frac{\partial f(x_1, x_2, x_3, x_4)}{\partial (x_2, x_4)}} \wedge \Delta_{x_2, x_4} f(x_1, x_2, x_3, x_4) = x_1 \overline{x}_3 (x_2 \oplus x_4). \tag{3.28}$$

Hence, static hazards with regard to a simultaneous change of x_2 and x_4 exist for the given function only for the input patterns $(x_1, x_2, x_3, x_4) = (1100)$ and $(x_1, x_2, x_3, x_4) = (1001)$. Figure 3.3 shows that these two patterns are static 0 - hazards which hold for the conditions given above.

Using a vectorial and a 2-fold maximum, the patterns of the static 0-hazards can be calculated separately by

$$\overline{\max_{(x_i, x_j)} f(x_i, x_j, \mathbf{x_1})} \wedge \max_{(x_i, x_j)}{}^2 f(x_i, x_j, \mathbf{x_1}) = 1, \tag{3.29}$$

while the vectorial and a 2-fold minimum describe, in the following equation, the static 1-hazards:

$$\min_{(x_i, x_j)} f(x_i, x_j, \mathbf{x_1}) \wedge \overline{\min_{(x_i, x_j)}{}^2 f(x_i, x_j, \mathbf{x_1})} = 1. \tag{3.30}$$

Dynamic hazards require that three variables of the function change their values. The comparison of the signal sequences of static and dynamic hazards shows that a dynamic hazard can be expressed by a static hazard with regard to two of the given variables followed by the change of the function caused by the third variable. Hence, for $f(\mathbf{x}) = f(x_i, x_j, x_k, \mathbf{x_1})$, the dynamic 0-hazards can be calculated by (3.31) and the dynamic 1-hazards by (3.32) with regard to the change of the variables (x_i, x_j, x_k):

$$\overline{\max_{(x_i, x_j)} f(\mathbf{x})} \wedge \max_{(x_i, x_j)}{}^2 f(\mathbf{x}) \wedge \frac{\partial f(\mathbf{x})}{\partial x_k} = 1, \tag{3.31}$$

$$\min_{(x_i, x_j)} f(\mathbf{x}) \wedge \overline{\min_{(x_i, x_j)}{}^2 f(\mathbf{x})} \wedge \frac{\partial f(\mathbf{x})}{\partial x_k} = 1. \tag{3.32}$$

Additionally, a dynamic hazard can be expressed by the change of the function caused by the variable x_k followed by a static hazard with regard to the other two variables (x_i, x_j). Hence, using exchanged associations dynamic 1-hazards can be calculated by (3.31) and the dynamic 0-hazards by (3.32), too.

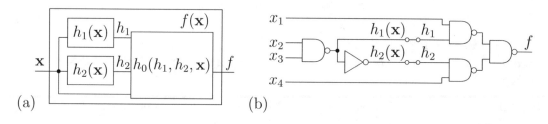

(a)   (b)

Figure 3.4: Structural hazards: (a) circuit model, (b) circuit example.

In order to find the conditions for **structural hazards**, we must apply the known condition of functional hazards to the real circuit structure. From (3.25), (3.29), and (3.30), we know that two variables of a function must change their value at the same time to cause a structural hazard. These two variables can be mapped to any pair of signals of a real circuit. Figure 3.4(a) shows the circuit model that allows to detect any hazard in a real circuit. The pair of signals that causes finally a hazard are labeled by h_1 and h_2. Basically, the hazard is caused by the change of the values $\mathbf{x}_0 \subseteq \mathbf{x}$ at the same time. The set of variables $\mathbf{x}_0$ must include at least one variable, any subset of variables of the function $f(\mathbf{x})$, or even all variables of this function.

The conditions of all *structural static hazards* of a circuit modeled by Figure 3.4(a), basically, with regard to $\mathbf{x}_0$ and finally with regard to h_1 and h_2 are:

1. There must be a static hazard for $h_0(h_1, h_2, \mathbf{x})$ with regard to h_1 and h_2.

2. The change of all variables of $\mathbf{x}_0$ at the same time must change both the function $h_1(\mathbf{x})$ and the function $h_2(\mathbf{x})$.

3. The change of all variables of $\mathbf{x}_0$ at the same time does not change the function $f(\mathbf{x})$.

Hence, all structural static 0-hazards explicitly caused by the change of $\mathbf{x}_0$ and implicitly caused by the change of both h_1 and h_2 can be calculated by:

$$\max_{(h_1,h_2)}{}^2 \left[\overline{\max_{(h_1,h_2)} h_0(\mathbf{h}, \mathbf{x})} \right] \wedge \max_{(h_1,h_2)}{}^2 h_0(\mathbf{h}, \mathbf{x}) \wedge \frac{\partial h_1(\mathbf{x})}{\partial \mathbf{x}_0} \wedge \frac{\partial h_2(\mathbf{x})}{\partial \mathbf{x}_0} \wedge \overline{\frac{\partial f(\mathbf{x})}{\partial \mathbf{x}_0}} = 1. \qquad (3.33)$$

while all respective structural static 1-hazards can be calculated by:

$$\max_{(h_1,h_2)}{}^2 \left[\min_{(h_1,h_2)} h_0(\mathbf{h}, \mathbf{x}) \right] \wedge \overline{\min_{(h_1,h_2)}{}^2 h_0(\mathbf{h}, \mathbf{x})} \wedge \frac{\partial h_1(\mathbf{x})}{\partial \mathbf{x}_0} \wedge \frac{\partial h_2(\mathbf{x})}{\partial \mathbf{x}_0} \wedge \overline{\frac{\partial f(\mathbf{x})}{\partial \mathbf{x}_0}} = 1. \qquad (3.34)$$

Example 3.10 Consider function $f(x_1, x_2, x_3, x_4)$ of Figure 3.4(b). Suppose that (x_2, x_3) change at the same time and structural static 1-hazards with regard to h_1 and h_2 must be found. Using (3.34), we get structural static 1-hazards for the input patterns $(x_1, x_2, x_3, x_4) = (1001)$ and $(x_1, x_2, x_3, x_4) = (1111)$.

Using the same formula (3.34) with regard to the change of the single variable $\mathbf{x}_0 = x_2$ and the same implicitly changed internal signals h_1 and h_2, we find structural static 1-hazards for $(x_1, x_2, x_3, x_4) = (1011)$ and $(x_1, x_2, x_3, x_4) = (1111)$.

A structural dynamic hazard combines a structural static hazard with the change of the function value caused by the change of $\mathbf{x}_0$. Hence, we have to remove the negation operation of the last term in (3.33) and (3.34) to get all structural dynamic hazards.

Using the circuit model of Figure 3.4 (a) that allows detection of any hazard in a real circuit and (3.25), we get a single formula for all structural dynamic hazards caused basically by $\mathbf{x}_0$ and implicitly by h_1 and h_2:

$$\max_{(h_1,h_2)}{}^2 \left[\frac{\partial h_0(\mathbf{h}, \mathbf{x})}{\partial (h_1, h_2)} \right] \wedge \Delta_{(h_1,h_2)} h_0(\mathbf{h}, \mathbf{x}) \wedge \frac{\partial h_1(\mathbf{x})}{\partial \mathbf{x}_0} \wedge \frac{\partial h_2(\mathbf{x})}{\partial \mathbf{x}_0} \wedge \frac{\partial f(\mathbf{x})}{\partial \mathbf{x}_0} = 1. \qquad (3.35)$$

It should be mentioned that the smallest set of variables $\mathbf{x}_0$ includes only a single variable that can cause both the change of the function $f(\mathbf{x})$ and a static hazard with regard to h_1 and h_2, where the change of h_1 and h_2 is caused by this single variable. An examples will be given as Exercise 3.5

3.7   APPLICATIONS TO DECOMPOSITION

The size of the circuit, in general, increases exponentially with the number of variables. Therefore, it is helpful to separate single variables from the function using an AND-, OR- or EXOR-gate at the output of the circuit if this is possible. Figure 3.5 shows the circuit structure for these three types of separations of a single variable. A separation is the simplest type of a functional decomposition.

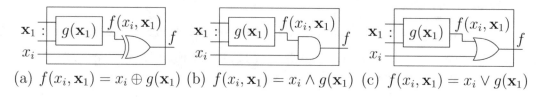

(a) $f(x_i, \mathbf{x}_1) = x_i \oplus g(\mathbf{x}_1)$ (b) $f(x_i, \mathbf{x}_1) = x_i \wedge g(\mathbf{x}_1)$ (c) $f(x_i, \mathbf{x}_1) = x_i \vee g(\mathbf{x}_1)$

Figure 3.5: Separation of a single variable x_i from a logic function $f(x_i, \mathbf{x}_1)$: (a) EXOR-separation, (b) AND-separation, (c) OR-separation.

The variable x_i can be separated from the logic function $f(x_i, \mathbf{x}_1)$ by an EXOR-gate if and only if

$$\frac{\partial f(x_i, \mathbf{x}_1)}{\partial x_i} = 1. \tag{3.36}$$

If (3.36) holds for any $\mathbf{x}_1$, the remainder function $g(\mathbf{x}_1)$ can be calculated by

$$g(\mathbf{x}_1) = \max_{x_i}(\overline{x}_i \wedge f(x_i, \mathbf{x}_1)) = f(x_i = 0, \mathbf{x}_1). \tag{3.37}$$

Alternatively, $\overline{x}_i$ can be separated from $f(x_i, \mathbf{x}_1)$ when (3.36) holds and the complement of $f(\mathbf{x}_1)$ (3.37) is used.

The variable x_i can be separated from the logic function $f(x_i, \mathbf{x}_1)$ by an AND-gate if and only if

$$\max_{x_i}(\overline{x}_i \wedge f(x_i, \mathbf{x}_1)) = f(x_i = 0, \mathbf{x}_1) = 0. \tag{3.38}$$

If (3.38) holds for any $\mathbf{x}_1$, the remainder function $g(\mathbf{x}_1)$ can be calculated by

$$g(\mathbf{x}_1) = \max_{x_i}(x_i \wedge f(x_i, \mathbf{x}_1)) = f(x_i = 1, \mathbf{x}_1). \tag{3.39}$$

The variable x_i can be separated from the logic function $f(x_i, \mathbf{x}_1)$ by an OR-gate if and only if

$$\overline{\min_{x_i}(\overline{x}_i \vee f(x_i, \mathbf{x}_1))} = \max_{x_i}(x_i \wedge \overline{f(x_i, \mathbf{x}_1)}) = \overline{f(x_i = 1, \mathbf{x}_1)} = 0. \tag{3.40}$$

If (3.40) holds for any $\mathbf{x}_1$, the remainder function $g(\mathbf{x}_1)$ can be calculated by

$$g(\mathbf{x}_1) = \max_{x_i}(\overline{x}_i \wedge f(x_i, \mathbf{x}_1)) = f(x_i = 0, \mathbf{x}_1). \tag{3.41}$$

Similar equations exist for the separation of $\overline{x}_i$ using an AND- or OR-gate. In this case, it is necessary to replace the single variables x_i or $\overline{x}_i$, respectively, by their complement (in the formulas (3.38), (3.39) (3.40), and (3.41)).

The separations shown in Figure 3.5 are helpful, but they rarely occur. More functions can be simplified using *bi-decompositions*. For a bi-decomposition, the variables $\mathbf{x}$ are split into three disjoint sets $\mathbf{x}_a, \mathbf{x}_b$, and $\mathbf{x}_c$, respectively. In a bi-decomposition, the function $f(\mathbf{x}_a, \mathbf{x}_b, \mathbf{x}_c)$ must be represented by the pair of functions $g(\mathbf{x}_a, \mathbf{x}_c)$ and $h(\mathbf{x}_b, \mathbf{x}_c)$, which only have the variables $\mathbf{x}_c$ in common. Three types of bi-decompositions exist:

- AND-bi-decomposition: $f(\mathbf{x}_a, \mathbf{x}_b, \mathbf{x}_c) = g(\mathbf{x}_a, \mathbf{x}_c) \wedge h(\mathbf{x}_b, \mathbf{x}_c),$ (3.42)
- OR-bi-decomposition: $f(\mathbf{x}_a, \mathbf{x}_b, \mathbf{x}_c) = g(\mathbf{x}_a, \mathbf{x}_c) \vee h(\mathbf{x}_b, \mathbf{x}_c),$ (3.43)
- EXOR-bi-decomposition: $f(\mathbf{x}_a, \mathbf{x}_b, \mathbf{x}_c) = g(\mathbf{x}_a, \mathbf{x}_c) \oplus h(\mathbf{x}_b, \mathbf{x}_c).$ (3.44)

Figure 3.6 shows the circuit structure for these three types of bi-decompositions of $f(\mathbf{x}_a, \mathbf{x}_b, \mathbf{x}_c)$ with regard to the sets of variables $\mathbf{x}_a$ and $\mathbf{x}_b$. We assume that $\mathbf{x}_a$ consists of k variables and $\mathbf{x}_b$ of l variables.

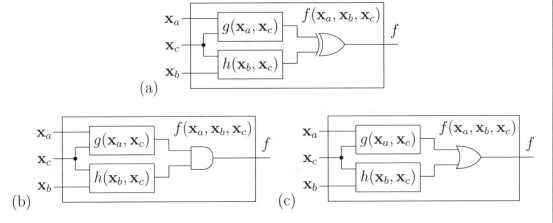

Figure 3.6: Circuit structure for the bi-decomposition of a function $f(\mathbf{x}_a, \mathbf{x}_b, \mathbf{x}_c)$ with regard to the sets of variables $\mathbf{x}_a$ and $\mathbf{x}_b$: (a) EXOR-bi-decomposition, (b) AND-bi-decomposition, (c) OR-bi-decomposition.

The logic function $f(\mathbf{x}_a, \mathbf{x}_b, \mathbf{x}_c)$ is AND-bi-decomposable with regard to the sets of variables $\mathbf{x}_a$ and $\mathbf{x}_b$ if and only if

$$\overline{f(\mathbf{x}_a, \mathbf{x}_b, \mathbf{x}_c)} \wedge \max_{\mathbf{x}_a}^{k} f(\mathbf{x}_a, \mathbf{x}_b, \mathbf{x}_c) \wedge \max_{\mathbf{x}_b}^{l} f(\mathbf{x}_a, \mathbf{x}_b, \mathbf{x}_c) = 0. \tag{3.45}$$

Formula (3.45) shows a necessary and sufficient condition for the AND-bi-decomposition: it is not allowed that a 0-value of the function is covered by the projection of 1-values in both the x_a-direction and the x_b-direction.

Generally, there are several different decomposition functions. One pair of decomposition functions of the AND-bi-decomposition is:

$$g(x_a, x_c) \;=\; \max_{x_b}^l f(x_a, x_b, x_c), \tag{3.46}$$

$$h(x_b, x_c) \;=\; \max_{x_a}^k f(x_a, x_b, x_c). \tag{3.47}$$

The validity of both the condition (3.45) and the calculation (3.46), (3.47) of the AND-bi-decomposition can be shown as follows: from (3.21), we have

$$f(x_a, x_b, x_c) \;\leq\; \max_{x_b}^l f(x_a, x_b, x_c), \tag{3.48}$$

$$f(x_a, x_b, x_c) \;\leq\; \max_{x_a}^k f(x_a, x_b, x_c), \tag{3.49}$$

$$f(x_a, x_b, x_c) \;\leq\; \max_{x_b}^l f(x_a, x_b, x_c) \wedge \max_{x_a}^k f(x_a, x_b, x_c). \tag{3.50}$$

If the condition (3.45) holds, then from (3.3), the inequality (3.50) can be transformed into the equation

$$f(x_a, x_b, x_c) = \max_{x_b}^l f(x_a, x_b, x_c) \wedge \max_{x_a}^k f(x_a, x_b, x_c). \tag{3.51}$$

Hence, under the condition that (3.45) holds, (3.42) follows directly from (3.51), (3.46), and (3.47).

Dual conditions and rules were found for the OR-bi-decomposition. The logic function $f(x_a, x_b, x_c)$ is OR-bi-decomposable with regard to the sets of variables x_a and x_b if and only if

$$f(x_a, x_b, x_c) \wedge \overline{\min_{x_a}^k f(x_a, x_b, x_c) \vee \min_{x_b}^l f(x_a, x_b, x_c)} = 0. \tag{3.52}$$

The condition (3.52) can be transformed into the equivalent equation

$$f(x_a, x_b, x_c) \wedge \max_{x_a}^k \overline{f(x_a, x_b, x_c)} \wedge \max_{x_b}^l \overline{f(x_a, x_b, x_c)} = 0. \tag{3.53}$$

Equation (3.53) can be expressed as follows: it is necessary and sufficient for an OR-bi-decomposition that no 1-value of the function is covered by the projection of 0-values in both the x_a-direction and the x_b-direction. One pair of decomposition functions of the OR-bi-decomposition can be calculated by

$$g(x_a, x_c) \;=\; \min_{x_b}^l f(x_a, x_b, x_c), \tag{3.54}$$

$$h(x_b, x_c) \;=\; \min_{x_a}^k f(x_a, x_b, x_c). \tag{3.55}$$

The similarity of the AND-bi-decomposition and the OR-bi-decomposition follows from the properties of the AND- and OR-operations. Both operations are nonlinear operations of the *Boolean Algebra*.

In contrast, the EXOR-operation is a linear operation of the algebraic structure of a *Boolean Ring*. Therefore, first, we consider the simplest case of the EXOR-bi-decomposition with regard to single variables a and b.

The logic function $f(a, b, \mathbf{x}_c)$ is EXOR-bi-decomposable with regard to the single variables a and b if and only if

$$\frac{\partial^2 f(a, b, \mathbf{x}_c)}{\partial a \, \partial b} = 0. \tag{3.56}$$

An even number of function values 1 in each subspace $(\mathbf{x}_c)$ allows the EXOR-bi-decomposition. The decomposition function $h(b, \mathbf{x}_c)$ can be calculated by

$$h(b, \mathbf{x}_c) = \max_a (a \wedge f(a, b, \mathbf{x}_c)), \tag{3.57}$$

which determines $g(a, \mathbf{x}_c)$ together with the basic formula of the EXOR-bi-decompositon (3.44) as follows:

$$g(a, \mathbf{x}_c) = f(a, b, \mathbf{x}_c) \oplus h(b, \mathbf{x}_c), \tag{3.58}$$
$$g(a, \mathbf{x}_c) = f(a, b, \mathbf{x}_c) \oplus \max_a (a \wedge f(a, b, \mathbf{x}_c)). \tag{3.59}$$

(3.59) reveals the condition of the EXOR-bi-decomposition: the expression on the left side of this equation (3.59) must not depend on b which can be expressed by

$$\frac{\partial g(a, \mathbf{x}_c)}{\partial b} = 0, \tag{3.60}$$
$$\frac{\partial [f(a, b, \mathbf{x}_c) \oplus \max_a (a \wedge f(a, b, \mathbf{x}_c))]}{\partial b} = 0. \tag{3.61}$$

Using Shannon's decomposition (3.1) and the definition of the simple maximum (3.6), the condition (3.61) can be transformed into the (already introduced) condition (3.56) of the EXOR-bi-decomposition with regard to the single variables a and b:

$$\frac{\partial [\overline{a} \cdot f(a = 0, b, \mathbf{x}_c) \oplus a \cdot f(a = 1, b, \mathbf{x}_c) \oplus f(a = 1, b, \mathbf{x}_c)]}{\partial b} = 0, \tag{3.62}$$
$$\frac{\partial [\overline{a} \cdot f(a = 0, b, \mathbf{x}_c) \oplus \overline{a} \cdot f(a = 1, b, \mathbf{x}_c)]}{\partial b} = 0, \tag{3.63}$$
$$\overline{a} \cdot \frac{\partial [f(a = 0, b, \mathbf{x}_c) \oplus f(a = 1, b, \mathbf{x}_c)]}{\partial b} = 0, \tag{3.64}$$
$$\frac{\partial}{\partial b} \left(\frac{\partial f(a, b, \mathbf{x}_c)}{\partial a} \right) = 0, \tag{3.65}$$
$$\frac{\partial^2 f(a, b, \mathbf{x}_c)}{\partial a \, \partial b} = 0. \tag{3.66}$$

Example 3.11   Consider function

$$f(a, b, x_1, x_2) = \bar{a}\,\bar{x}_1\,x_2 \vee a\,\bar{b}\,x_1\,x_2 \vee \bar{a}\,b\,x_1 \vee \bar{a}\,x_1\,\bar{x}_2 \tag{3.67}$$

shown in Figure 3.7(a). Each subfunction $f(a, b, \mathbf{x} = \mathbf{c})$ possesses an even number of function values 1. Hence, condition (3.56) for the EXOR-bi-decomposition with regard to a and b holds. As result of (3.57) and (3.59), we get the decomposition functions:

$$h(b, x_1, x_2) = \bar{b}\,x_1\,x_2. \tag{3.68}$$
$$g(a, x_1, x_2) = \bar{a}\,(x_1 \vee x_2). \tag{3.69}$$

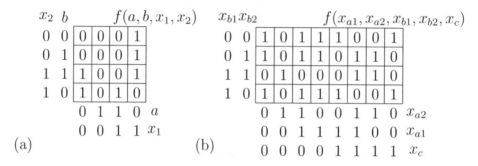

(a)   (b)

Figure 3.7: Functions as example for EXOR-bi-decomposition: (a) with regard to a and b, (b) with regard to (x_{a1}, x_{a2}) and (x_{b1}, x_{b2}).

Formula (3.56) can be generalized from a single variable b to a set of variables $\mathbf{x}_b$. The logic function $f(a, \mathbf{x}_b, \mathbf{x}_c)$ is EXOR-bi-decomposable with regard to the single variable a and a set of variables $\mathbf{x}_b$ if and only if

$$\Delta_{\mathbf{x}_b}\left(\frac{\partial f(a, \mathbf{x}_b, \mathbf{x}_c)}{\partial a}\right) = 0. \tag{3.70}$$

This more general EXOR-bi-decomposition requires that in a subspace $(\mathbf{x}_c)$ the change of the function values with regard to a must be the same for each $\mathbf{x}_b$.

The most general EXOR-bi-decomposition of a completely specified logic function with regard to the two sets of variables $\mathbf{x}_a$ and $\mathbf{x}_b$ combines the construction of a possible solution with a check of whether such an EXOR-bi-decomposition exists.

Let (3.71) be the potential EXOR-decomposition function $g(\mathbf{x}_a, \mathbf{x}_c)$ of the logic function $f(\mathbf{x}_a, \mathbf{x}_b, \mathbf{x}_c)$:

$$g(\mathbf{x}_a, \mathbf{x}_c) = f(\mathbf{x}_a, \mathbf{x}_b = (0, 0, \ldots, 0), \mathbf{x}_c). \tag{3.71}$$

Then, $f(\mathbf{x}_a, \mathbf{x}_b, \mathbf{x}_c)$ is EXOR-bi-decomposable with regard to the sets of variables $\mathbf{x}_a$ and $\mathbf{x}_b$ if and only if

$$\Delta_{\mathbf{x}_a}\left(f(\mathbf{x}_a, \mathbf{x}_b, \mathbf{x}_c) \oplus f(\mathbf{x}_a, \mathbf{x}_b = (0, 0, \ldots, 0), \mathbf{x}_c)\right) = 0. \tag{3.72}$$

If (3.72) holds, then (3.71) and

$$h(\mathbf{x}_b, \mathbf{x}_c) = \max_{\mathbf{x}_a}^k (f(\mathbf{x}_a, \mathbf{x}_b, \mathbf{x}_c) \oplus f(\mathbf{x}_a, \mathbf{x}_b = (0, 0, \dots, 0), \mathbf{x}_c)) \tag{3.73}$$

are possible decomposition functions of the EXOR-bi-decomposition of the function $f(\mathbf{x}_a, \mathbf{x}_b, \mathbf{x}_c)$ with regard to the sets of variables $\mathbf{x}_a$ and $\mathbf{x}_b$.

Example 3.12   Consider function

$$f(\mathbf{x}_a, \mathbf{x}_b, x_c) = \overline{x}_{a2} \overline{x}_{b1} \overline{x}_c \vee x_{a1} \overline{x}_{b1} \overline{x}_c \vee \overline{x}_{a2} \overline{x}_{b2} \vee \overline{x}_{a1} x_{a2} x_{b1} x_{b2} \vee x_{a2} x_{b2} x_c \tag{3.74}$$

shown in Figure 3.7(b). From (3.71), a potential EXOR-decomposition function of $f(\mathbf{x}_a, \mathbf{x}_b, x_c)$ with regard to (x_{a1}, x_{a2}) and (x_{b1}, x_{b2}) is

$$g(\mathbf{x}_a, x_c) = x_{a1} \overline{x}_c \vee \overline{x}_{a2}, \tag{3.75}$$

The condition (3.72) holds, so that (3.75) and, from (3.73), the function

$$h(\mathbf{x}_b, x_c) = x_{b2} \wedge (x_{b1} \vee x_c) \tag{3.76}$$

are allowed EXOR-decomposition functions of the given function $f(\mathbf{x}_a, \mathbf{x}_b, x_c)$.

The conditions for the bi-decomposition also restrict the number of logic functions for which such a decomposition exists. Taking incompletely specified logic functions into account, the chance to find a bi-decomposition grows as the number of don't-care values increases. Completeness of the bi-decomposition means that for each function at least one bi-decomposition exists. The completeness of the bi-decomposition is reached and could be proven using the BDC, when the weak bi-decomposition (with $\mathbf{x}_b = \emptyset$) is used together with the strong bi-decomposition, which is studied in this section. These interesting extensions are explained in [11] and [16].

3.8   APPLICATION TO TEST PATTERN GENERATION

The stuck-at fault model is simple and frequently used. There are two types of stuck-at faults. If a circuit fault causes a constant value 0, then it is a stuck-at-zero fault (s-a-0). Otherwise, if a fault causes a constant value 1, it is a stuck-at-one fault (s-a-1). We introduce an additional logic variable T into the model, to express an s-a-T-fault:

$$T = \begin{cases} 0, & \text{if a stuck-at-0 fault } (s\text{-}a\text{-}0) \text{ is considered,} \\ 1, & \text{if a stuck-at-1 fault } (s\text{-}a\text{-}1) \text{ is considered.} \end{cases} \tag{3.77}$$

To test a combinational circuit, test patterns are necessary. Typically, access is restricted to the inputs and outputs of the circuit. Thus, it is difficult to find test patterns that check an internal wire. In order to generate test patterns, we need the function $g(\mathbf{x})$ that controls this wire, and the function $h(\mathbf{x}, s)$ where s is the signal of the wire to be tested.

Three conditions must hold for a test pattern to detect an s-a-T-fault.

1. It is necessary to control the circuit in such a way that the function value $s = g(\mathbf{x})$ is the complement of the stuck-at-T-fault to be checked at the wire s. Only in this case can an s-a-T-fault cause a change of the circuit output. The solutions of the equation (3.78) enable this effect of a s-a-T-fault. Therefore, we call the equation (3.78) *fault controllability condition*:

$$g(\mathbf{x}) \oplus T = 1. \tag{3.78}$$

2. We assume that only the output of the circuit is directly observable. The effect of the internal fault at the wire s must influence the function value at the circuit output. Thus, a further necessary condition for test patterns is the *fault observability*. The value at the wire s is observable by the value at the circuit output for all solution vectors $\mathbf{x}$ of (3.79):

$$\frac{\partial h(\mathbf{x}, s)}{\partial s} = 1. \tag{3.79}$$

3. It is necessary to compare the observed output value y with the expected function value of $f(\mathbf{x})$ to evaluate whether there exists an s-a-T-fault at the wire s or not. The expected value y is related to the input pattern $\mathbf{x}$ in the solution of the logic equation (3.80). The solution of the equation (3.80) allows the *fault evaluation*:

$$f(\mathbf{x}) \odot y = 1, \tag{3.80}$$

where $a \odot b = \bar{a} \oplus b$.

The three conditions (3.78), (3.79), and (3.80) are collectively sufficient and can be combined to build the necessary and sufficient condition (3.81). The solutions of the equation (3.81) comprise the set of test patterns $(\mathbf{x}, y, T)$. Each of these patterns detects an s-a-T-fault if the circuit does not show the value y controlled by the input $\mathbf{x}$:

$$(g(\mathbf{x}) \oplus T) \wedge \frac{\partial h(\mathbf{x}, s)}{\partial s} \wedge (f(\mathbf{x}) \odot (y) = 1. \tag{3.81}$$

BDC allows the simplification of test pattern generation [11], and allows an extension to both internal branches [11] and bridging faults [4]. It should be mentioned that the bi-decomposition introduced before allows us to compute circuits that are completely testable with regard to s-a-T-faults at all inputs, outputs and gate-connections [11]. These multi-level circuits have a very small depth [10]. Furthermore, these test patterns can be generated in parallel to synthesize the circuit [17], so that only a small overhead of about 10% solves the test pattern generation.

3.9   GENERALIZATION TO DIFFERENTIAL OPERATIONS

As an extension to the derivative operations, the differentials allow the specification of the direction of change, explicitly.

Definition 3.13   Let x_i be a Boolean variable, then

$$\mathrm{d}x_i = \begin{cases} 1, & \text{if the variable } x_i \text{ changes its value} \\ 0, & \text{if the variable } x_i \text{ does not change its value} \end{cases} \tag{3.82}$$

is the differential of the Boolean variable x_i.

Equations that contain variables x_i and differentials $\mathrm{d}x_i$ will be used mainly to model graphs and their properties. Similar to the change of a variable, the change of a function can be defined.

Definition 3.14   Let $f(\mathbf{x})$ be a logic function of n variables. Then

$$\mathrm{d}_{\mathbf{x}} f(\mathbf{x}) = f(\mathbf{x}) \oplus f(\mathbf{x} \oplus \mathrm{d}\mathbf{x}) \tag{3.83}$$

is the (total) differential of the function $f(\mathbf{x})$ with regard to all variables of $\mathbf{x}$.

Replacing the EXOR operation between the functions in (3.83) by AND or OR, the (total) differential minimum or (total) differential maximum are defined. The restriction to a single variable in these definitions leads to three partial differential operations. Their iterative execution for several variables leads to three m-fold differential operations. Due to the available space, we skip the detail definition of all these differential operations and refer to [11]. It should be mentioned that each differential operation expresses several appropriated derivative operations associated to the direction of change.

3.10   SUMMARY

This chapter gives an introduction into Boolean Differential Calculus (BDC). BDC defines derivative operations as well as differential operations to describe different types of changes. This chapter explores different types of derivative operations, introduces some of the basic theorems and explains some typical applications for digital circuits.

In order to apply BDC to real-world problems, it is highly recommended to use the software system XBOOLE since the operations have been directly implemented.

3.11   EXERCISES

3.1. Prove the second inequality of Theorem 3.2 similarly to the proof of the first inequality of the same theorem.

3.2. Prove that the relations given in (3.7), (3.8) and (3.9) are equivalent to Definition 3.1.

3.3. Prove both inequalities of Theorem 3.5 similarly to the proof given for Theorem 3.2.

3.4. Prove the rightmost inequality of Theorem 3.7 similarly to the proof of the leftmost inequality of the same theorem.

3.5. Calculate the structural dynamic hazards of the circuit which are explicitly caused be the changes of (a) x_2, (b) x_3, and (c) (x_2, x_3), where the embedded static hazard are caused by the signals h_1 and h_2. How can the circuit be changed to avoid the detected hazards? Verify that the primarily found structural dynamic hazards are removed by the added gate.

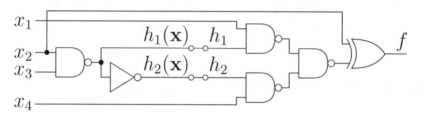

We suggest the use of the XBOOLE monitor that can be downloaded for free from the WEB page http://www.informatik.tu-freiberg.de/xboole. The integrated help system describes how to use this powerful tool. Because six similar tasks must be solved, it is suggested to prepare and apply an XBOOLE problem program (PRP).

3.6. Show that OR-bi-decomposition (3.43) can be realized by the decomposition functions (3.54) and (3.55) if condition (3.52) holds.

3.7. Specify a logic function $h(\mathbf{x}_b, \mathbf{x}_c)$ usable in an EXOR-bi-decomposition of $f(a, \mathbf{x}_b, \mathbf{x}_c)$ with regard to the single variable a and the set of variables $\mathbf{x}_b$.

3.8. Specify a logic function $g(a, \mathbf{x}_c)$ that can be used together with $h(\mathbf{x}_b, \mathbf{x}_c)$ of Exercise 3.7 in an EXOR-bi-decomposition of $f(a, \mathbf{x}_b, \mathbf{x}_c)$.

3.9. Verify that the functions chosen in Exercises 3.7 and 3.8 satisfy the condition (3.70) of an EXOR-bi-decomposition of $f(a, \mathbf{x}_b, \mathbf{x}_c)$ with regard to the single variable a and the set of variables $\mathbf{x}_b$.

3.10. Find all pairs of variables that allow an EXOR-bi-decomposition for the function $f(\mathbf{x}) = (((x_1 \oplus x_2) \wedge x_3) \vee x_4) \oplus ((x_2 \wedge x_5) \vee (x_4 \oplus x_6))$. Calculate the decomposition functions $g(\mathbf{x}_a, \mathbf{x}_c)$ and $h(\mathbf{x}_b, \mathbf{x}_c)$ having the smallest common set of variables. Verify your result of the EXOR -bi-decomposition. We suggest the use of the XBOOLE monitor.

3.11. Continue the bi-decomposition of both decomposition functions found in Exercises 3.10 and draw the circuit structure using gates of two inputs only.

REFERENCES

[1] J. S. B. Akers, "On a theory of Boolean functions," *Journal of the Society for Industrial and Applied Mathematics (SIAM)*, 7(4):487–498, 1959. DOI: 10.1137/0107041 3.1

[2] D. Bochmann, *Binary Systems - A BOOLEAN Book*, TUDpress, Verlag der Wissenschaften GmbH, Dresden, Germany, 2008. 3.1

[3] D. Bochmann, and C. Posthoff, *Binäre Dynamische Systeme*, Akademie-Verlag, Berlin, 1981. 3.1, 3.5

[4] I. I. Bucur, M. Drăgoicea, and N. Constantin, "Boolean differential calculus applied in logic testing," *Studies in Informatics and Control, with Emphasis on Useful Applications of Advanced Technology*, 16(2), ISSN 1220-1766, 2007. 3.1, 3.8

[5] D. A. Huffman, "Solvability criterion for simultaneous logical equations," *Quarterly Progress Report*, 1(56):87–88, 1958. 3.1

[6] S. Kolodzinski, and E. Hrynkiewicz, "An utilisation of Boolean differential calculus in variables partition calculation for decomposition of logic functions," *Design and Diagnostics of Electronic Circuits and Systems*, Los Alamitos, CA, USA, ISBN 978-1-4244-3341-4, pp. 34–37, 2009. DOI: 10.1109/DDECS.2009.5012095 3.1

[7] C. Lang and B. Steinbach, "Bi-decomposition of function sets in multiple-valued logic for circuit design and data mining," *S. N. Yanushkevich: Artificial Intelligence in Logic Design*, Kluwer Academic Publisher, Dordrecht, The Netherlands, pp. 73–107, 2004. 3.1

[8] I. S. Reed, "A class of multiple-error-correcting codes and the decoding scheme," *Transactions of the IRE Professional Group on Information Theory*, 4(4):38–49, 1954. DOI: 10.1109/TIT.1954.1057465 3.1

[9] M. Davio, J.-P. Deschamps, and A. Thayse, *Discrete and Switching Functions*, McGraw-Hill International, 1978. 3.1

[10] A. Mishchenko, B. Steinbach, and M. Perkowski, "An algorithm for bi-decomposition of logic functions," *Proceedings of the 38th Design Automation Conference*, Las Vegas (Nevada) USA, pp. 103–108, 2001. DOI: 10.1145/378239.378353 3.1, 3.8

[11] C. Posthoff and B. Steinbach, *Logic Functions and Equations — Binary Models for Computer Science*, Springer, Dordrecht, The Netherlands, 2004. 3.1, 3.5, 3.7, 3.8, 3.9

[12] B. Steinbach, "XBOOLE - a toolbox for modelling, simulation, and analysis of large digital systems," *System Analysis and Modelling Simulation*, 9(4):297–312, 1992. 3.1

[13] B. Steinbach and C. Lang, "Complete bi-decomposition of multiple-valued functions using MIN and MAX gates," *Proceedings of the 35th International Symposium on Multiple-Valued Logic, (ISMVL 2005)*, Calgary, Canada, pp. 69–74, 2005. DOI: 10.1109/ISMVL.2005.14 3.1

[14] B. Steinbach and C. Lang, "Exploiting functional properties of Boolean functions for optimal multi-level design by bi-decomposition," *S. N. Yanushkevich: Artificial Intelligence in Logic Design*, Kluwer Academic Publisher, Dordrecht, The Netherlands, pp. 159-200, 2004. 3.1

[15] B. Steinbach, and C. Posthoff, "Extended theory of Boolean normal forms," *Proceedings of the 6th Annual Hawaii International Conference on Statistics, Mathematics and Related Fields*, Honolulu, Hawaii, pp. 1124–1139, 2007. 3.1

[16] B. Steinbach, and C. Posthoff, *Logic Functions and Equations - Examples and Exercises*. Springer Science + Business Media B.V., 2009. 3.1, 3.5, 3.7

[17] B. Steinbach, and M. Stöckert, "Design of fully testable circuits by functional decomposition and implicit test pattern generation," *Proceedings of the 12th IEEE VLSI Test Symposium*, Cherry Hill (New Jersey) USA, pp. 22–27, 1994. 3.1, 3.8

[18] A. Thayse, *Boolean Calculus of Differences*, Springer, Berlin, 1981. 3.1

[19] S. N. Yanushkevich, *Logic Differential Calculus in Multi-Valued Logic Design*, Habilitation thesis - ISBN 83-87423-16-5, Tech. University of Szczecin, Szczecin, Poland, 1998. 3.1

[20] S. N. Yanushkevich, V. P. Shmerko, and B. Steinbach, "Spatial interconnect analysis for predictable nanotechnologies," *Journal of Computational and Theoretical Nanoscience*, American Scientific Publishers, USA, 5(1), pp. 56–69, 2008. 3.1

CHAPTER 4

Synthesis of Boolean Functions in Reversible Logic

Robert Wille and Rolf Drechsler

CHAPTER SUMMARY

Traditional technologies like CMOS suffer more and more from the increasing miniaturization and the exponential growth of the number of transistors. Thus, alternatives that replace or at least enhance traditional computer chips are needed. Reversible logic, and its applications in domains like quantum computation and low-power design, is such a possible alternative and, thus, has become an intensely studied topic in the recent years. But, synthesis of reversible circuits significantly differs from traditional logic synthesis. In particular, fan-out and feedback are not allowed so that reversible circuits must be cascades of reversible gates. This requires completely new synthesis approaches.

      This chapter provides an introduction into the topic as well as an overview of selected synthesis methods for reversible logic. More precisely, we review reversible functions as well as reversible circuits and, in particular, focus on the embedding of irreversible functions into reversible ones. Then, we describe how such functions (given as a truth table) can be synthesized using exact as well as heuristic approaches. Since only small functions can be synthesized using a truth table as input, we describe a method that exploits *Binary Decision Diagrams* (BDDs) for reversible logic synthesis of significantly larger functions.

4.1   INTRODUCTION

Reversible logic [15, 2, 32] realizes n-input n-output functions that map each possible input vector to a unique output vector (i.e., bijections). Recently, synthesis of reversible logic has become an intensely studied research area. In particular, this is caused by the fact that reversible logic is the basis for several emerging technologies while traditional methods suffer from the increasing miniaturization and the exponential growth of the number of transistors in integrated circuits. Researchers expect that in 10-20 years duplication of transistor density every 18 months (according to *Moore's Law*) is not possible any longer (see e.g., [39]). Then, alternatives are needed. Reversible logic offers such alternatives as the following applications show:

- *Reversible Logic for Low-Power Design*
Power dissipation and therewith heat generation is a serious problem for today's computer chips. Landauer and Bennett showed in [15, 2] that 1. using traditional (irreversible) logic gates always leads to energy dissipation regardless of the underlying technology and 2. that circuits with zero power dissipation must be information-lossless. This holds for reversible logic, since data is bijectively transformed without losing any of the original information. Even if today energy dissipation is mainly caused by non-ideal behaviors of transistors and materials, the theoretically possible zero power dissipation makes reversible logic quite interesting for the future. Moreover, in 2002 first reversible circuits have been physically implemented [6] that exploit these observations in the sense that they are powered by their input signals only and did not need additional power supplies.

- *Reversible Logic as Basis for Quantum Computation*
Quantum circuits [23] offer a new kind of computation. Here, qubits instead of traditional bits are used that allow to represent not only 0 and 1 but also a superposition of both. As a result, qubits can represent multiple states at the same time enabling enormous speed-ups in computations. Even if research in the domain of quantum circuits is still at the beginning, first, quantum circuits have already been built. Reversible logic is important in this area because every quantum operation is inherently reversible. Thus, progress in the domain of reversible logic can directly be applied to quantum logic.

Besides that, further applications of reversible logic can be found in the domain of optical computing [5], DNA computing [2], and nanotechnologies [20]. However, reversible logic synthesis significantly differs from traditional approaches. This is because fan-out and feedback are not allowed and, thus, a circuit must be a cascade of reversible gates [23].

In this chapter, we review the basic concepts of reversible logic and describe current synthesis approaches that work on the function to be synthesized given as a truth table. More precisely, we

- show how an irreversible function can be embedded in a reversible specification,

- explain how an optimal realization for the given function can be obtained using an exact synthesis method, and

- describe how the function can be synthesized using a heuristic approach.

Moreover, since these synthesis approaches rely on a truth table (and therefore are only applicable to very small functions), we also describe a method that exploits *Binary Decision Diagrams* (BDDs) to synthesize significantly larger functions. Therewith, an overview of the most current synthesis approaches for reversible logic is given.

The remaining chapter is structured as follows: Section 4.2 briefly introduces reversible logic, the most commonly used gates, and the applied cost metrics. Afterwards, Section 4.3 describes the necessary steps to embed an irreversible function in a reversible specification. The synthesis using an exact and a heuristic approach (with a truth table as input) is described in Section 4.4 and Section 4.5, respectively. BDD-based synthesis for large functions is described in Section 4.6. Finally, the chapter is concluded and future work is sketched in Section 4.7.

4.2   REVERSIBLE LOGIC

Reversible logic realizes n-input n-output functions that map each possible input vector to a unique output vector – in other words bijections are realized. In reversible logic, fan-out and feedback are not allowed [23]. As a consequence, reversible circuits are cascades of reversible gates.

Let $X := \{x_1, \ldots, x_n\}$ be the set of Boolean variables. Then, a reversible gate has the form $g(C, T)$, where $C = \{x_{i_1}, \ldots, x_{i_k}\} \subset X$ is the set of *control lines* and $T = \{x_{j_1}, \ldots, x_{j_l}\} \subset X$ with $C \cap T = \emptyset$ is the set of *target lines*. In the past, multiple control Toffoli [32], multiple control Fredkin [9], and Peres [25] gates have been established:

- A *multiple control Toffoli gate* (TOF) has a single target line x_j. The gate maps $(x_1, x_2, \ldots, x_j, \ldots, x_n)$ to $(x_1, x_2, \ldots, x_{i_1} x_{i_2} \cdots x_{i_k} \oplus x_j, \ldots, x_n)$, i.e., the value of the target line is inverted if all control lines evaluate to 1.

- A *multiple control Fredkin gate* has two target lines x_{j_1} and x_{j_2}. The values of the target lines are interchanged if all control lines evaluate to 1.

- A *Peres gate* has one control line x_i and two target lines x_{j_1} and x_{j_2}. The gate maps $(x_1, x_2, \ldots, x_{j_1}, \ldots, x_{j_2} \ldots, x_n)$ to $(x_1, x_2, \ldots, x_i \oplus x_{j_1}, \ldots, x_i x_{j_1} \oplus x_{j_2}, \ldots, x_n)$, i.e., the Peres gate is a cascade of two Toffoli gates (with two and one control lines, respectively).

Figure 4.1 shows a Toffoli gate, a Fredkin gate, and a Peres gate in a cascade. The control lines are denoted by ●, while the target lines are denoted by ⊕ (except for the Fredkin gate where a × is used instead). The annotated values demonstrate the computation of the respective gates. As shown, the calculation can be done in both directions.

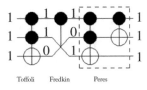

Toffoli   Fredkin   Peres

Figure 4.1: Reversible gates.

The *cost* of a reversible circuit is often defined by the number of gates. Besides that, *quantum cost* are also used. Quantum cost denotes the effort needed to realize a gate in quantum logic. More precisely, every reversible circuit can be transformed into a sequence of elementary quantum gates (for more details see e.g., [23]). Quantum cost gives the number of these quantum gates needed to realize a reversible gate. Table 4.1 shows the quantum cost for a some Toffoli and Fredkin gate configurations as introduced in [1] and optimized in [16] and [19][1]. For example, a Toffoli gate with two controls has a cost of five and a Fredkin gate with one control has a cost of seven. Additionally, a Peres gate has a cost of four. The Peres gate is of interest, since the realization with two Toffoli gates would imply a cost of six. The sum of the quantum cost for each gate defines the quantum cost for the whole circuit.

In the following, we focus on synthesis using Toffoli gates only.

Table 4.1: Quantum cost for Toffoli and Fredkin gates.

No. of control lines	Quantum cost	
	of a Toffoli gate	of a Fredkin gate
0	1	3
1	1	7
2	5	15
3	13	28, if at least 2 lines are unconnected 31, otherwise
4	26, if at least 2 lines are unconnected 29, otherwise	40, if at least 3 lines are unconnected 54, if 1 or 2 lines are unconnected 63, otherwise
5	38, if at least 3 lines are unconnected 52, if 1 or 2 lines are unconnected 61, otherwise	52, if at least 4 lines are unconnected 82, if 1, 2 or 3 lines are unconnected 127, otherwise

4.3   EMBEDDING IRREVERSIBLE FUNCTIONS

Table 4.2 shows the truth table of an 1-bit adder which is used as an example in the next sections. The adder has three inputs (the carry-bit c_{in} as well as the two summands x and y) and two outputs (the carry c_{out} and the *sum*). The adder obviously is irreversible since

- the number of inputs differs from the number of outputs and

- there is no unique input-output mapping.

[1]It should be noted that the definition of quantum cost is still subject to modifications, since constantly better decompositions of reversible gates to quantum circuits are obtained. Furthermore, the real cost obviously depend on the concrete technology. Thus, modifications in quantum cost may occur in the future.

Table 4.2: Truth table of an adder.

c_{in}	x	y	c_{out}	sum	
0	0	0	0	0	0
0	0	1	**0**	**1**	**0**
0	1	0	**0**	**1**	**1**
0	1	1	*1*	*0*	*0*
1	0	0	**0**	**1**	?
1	0	1	*1*	*0*	*1*
1	1	0	*1*	*0*	*?*
1	1	1	1	1	1

Even adding an additional output to the function (leading to the same number of input and outputs) would not make the function reversible. Then, the first four lines of the truth table can be embedded with respect to reversibility as shown in the rightmost column of Table 4.2. However, since $c_{out} = 0$ and $sum = 1$ already appeared two times (marked bold), no unique embedding for the fifth truth table line is possible any longer. The same also holds for the lines marked italic.

This already has been observed in [17]. There, the authors came to the conclusion that at least $\lceil log(m) \rceil$ free outputs are required to make an irreversible function reversible, where m is the maximum number of times an output pattern is repeated in the truth table. Since for the adder at most 3 output pattern are repeated, $\lceil log(3) \rceil = 2$ free outputs (and therewith one additional circuit line) are required to make the function reversible.

Adding new lines causes constant inputs and garbage outputs. The value of the constant inputs can be chosen by the designer. Garbage outputs are by definition don't cares and thus can be left unspecified leading to an incompletely specified function. However, many synthesis approaches require a completely specified function so that all don't cares must be assigned with a concrete value.

As a result, the adder is embedded in a reversible function including four variables, one constant input, and two garbage outputs. A possible assignment to the constant as well as the don't care values is depicted in Table 4.3. Note that the concrete embedding may influence the respective synthesis results. Corresponding evaluations have been made e.g., in [36, 22]. In the remainder of the chapter, we use the embedding of Table 4.3 to describe the synthesis approaches.

4.4   EXACT SYNTHESIS

Exact synthesis algorithms determine a *minimal circuit realization* for a given function, i.e., a circuit with a minimal number of gates or quantum cost, respectively. Ensuring minimality often causes a large computation time and thus exact approaches are only applicable to relatively small functions. Nevertheless, it is worth to consider exact methods since

- they allow finding smaller circuits than the currently best known realizations,

Table 4.3: Truth table of an embedded adder.

0	c_{in}	x	y	c_{out}	sum	g_1	g_2
0	0	0	0	0	0	0	0
0	0	0	1	0	1	1	1
0	0	1	0	0	1	1	0
0	0	1	1	1	0	0	1
0	1	0	0	0	1	0	0
0	1	0	1	1	0	1	1
0	1	1	0	1	0	1	0
0	1	1	1	1	1	0	1
1	0	0	0	1	0	0	0
1	0	0	1	1	1	1	1
1	0	1	0	1	1	1	0
1	0	1	1	0	0	0	1
1	1	0	0	1	1	0	0
1	1	0	1	0	0	1	1
1	1	1	0	0	0	1	0
1	1	1	1	0	1	0	1

- they allow the evaluation of the quality of heuristic approaches, and

- they allow the computation of minimal circuits as basic blocks for larger circuits.

For example, improving the heuristic results by 10% is significant, if optimal results are obtained, but marginal if the generated results are still factors away from the optimum. To obtain such conclusions, the optimum must be available.

As a result, in the past approaches synthesizing circuits with optimal quantum cost [13, 10] as well as optimal number of reversible gates [30, 11] have been proposed. In this section, we describe exact synthesis based on the method of [11]. Here, circuits with minimal number of Toffoli gates are obtained.

In [11], the exact synthesis problem is formulated as a sequence of decision problems. Given the reversible function $f : \mathbb{B}^n \to \mathbb{B}^n$, it is checked if f can be synthesized using $d = 1$ Toffoli gates. If this fails, then d is increased until a realization has been determined. Since d is iteratively increased starting with $d = 1$, minimality is ensured. The respective checks are thereby performed by

- encoding the synthesis problem as an instance of *Boolean satisfiability* (SAT) and

- using a SAT solver (e.g., [7]) to solve this instance.

In the following, we describe the concrete SAT encoding for the adder given in Table 4.3 and $d = 4$. The goal is to formulate an instance (i.e., a Boolean function) which becomes satisfiable (i.e., evaluates to 1) if a circuit with $d = 4$ gates representing the function exists. To this end, Boolean variables and constraints are introduced.

First, variables $x_{i0}^k, x_{i1}^k, \ldots, x_{in-1}^k$ are used to represent the input- (for $k = 0$), the output- (for $k = d$), and the auxiliary values (for $1 \leq k \leq d - 1$) of the circuit to be synthesized for each truth table line i of f. Therefore, the left side of the truth table corresponds to the variables $x_{i0}^0, x_{i1}^0, \ldots, x_{in-1}^0$, while the right side corresponds to the variables $x_{i0}^d, x_{i1}^d, \ldots, x_{in-1}^d$, respectively. Figure 4.2 shows the respective variables for the adder function given in Table 4.3. The first row of Figure 4.2 represents the variables for the first truth table line, the second row the ones for the second truth table line, and so on.

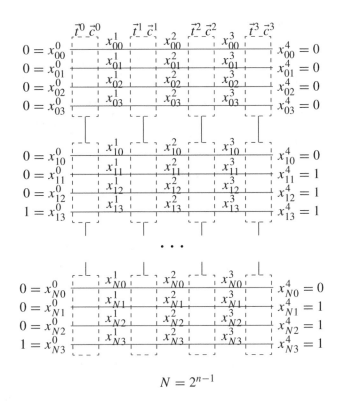

$$N = 2^{n-1}$$

Figure 4.2: Exact SAT formulation for $d = 4$.

Furthermore, variables $t_{\lceil \log_2 n \rceil}^k, t_{\lceil \log_2 n \rceil - 1}^k, \ldots, t_1^k$ and $c_1^k, c_2^k, \ldots, c_{n-1}^k$ with $0 \leq k < d$ are introduced, whose assignments represent the type of the Toffoli gate at depth k (for brevity denoted by $\vec{t}^k$ and $\vec{c}^k$ in the following). The variable $\vec{t}^k$ is thereby used as a binary encoding of a natural number $t^k \in \{0, \ldots, n - 1\}$ which defines the chosen target line. In contrast, $\vec{c}^k$ denotes the control lines. More precisely, assigning $c_l^k = 1$ ($1 \leq l \leq n - 1$) means that line $(t^k + l) \mod n$ becomes

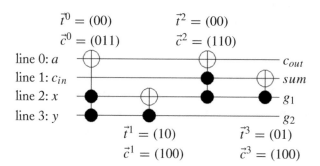

Figure 4.3: Circuit obtained by exact synthesis.

a control line of the Toffoli gate at depth k. Figure 4.3 gives some examples for assignments to $\vec{t}^k$ and $\vec{c}^k$ with their respective Toffoli gate representation.

　　Using these variables, constraints are introduced assigning the input and output of the truth table to their respective x_{ij}^0 and x_{ij}^d variables (see left and right side of Figure 4.2). Furthermore, for each gate to be synthesized at depth k, functional constraints are added so that (depending on the assignment to $\vec{t}^k$ and $\vec{c}^k$ as well as to the input x_{ij}^k) the respective gate output x_{ij}^{k+1} is computed. As an example, consider $\vec{t}^k = (01)$ and $\vec{c}^k = (001)$, i.e., with $c_3^k = 1$. This assignment states that the Toffoli gate at depth k has line $t^k = [01]_2 = 1$ as target line and line $(t^k + l) \mod n = (1 + 3) \mod 4 = 0$ as single control line. To cover this case, constraints

$$\vec{t}^k = (01) \wedge \vec{c}^k = (001) \Rightarrow \begin{aligned} & x_{i0}^{k+1} == x_{i0}^k \\ \wedge\ & x_{i1}^{k+1} == x_{i1}^k \oplus x_{i0}^k \\ \wedge\ & x_{i2}^{k+1} == x_{i2}^k \\ \wedge\ & x_{i3}^{k+1} == x_{i3}^k \end{aligned}$$

are added for each truth table line i. In other words, the values of lines 0, 2, and 3 are passed through, while the output value of line 1 becomes inverted, if line 0 is assigned to 1. Similar constraints are added for all remaining cases.

　　As a result, a functional description has been constructed which is satisfiable if there is a valid assignment to $\vec{t}^k$ and $\vec{c}^k$ so that for all truth table lines the desired input-output mapping is achieved. Then, the concrete Toffoli gates can be obtained by the assignments to $\vec{t}^k$ and $\vec{c}^k$ as depicted in Figure 4.3. If there is no such assignment (i.e., the instance is unsatisfiable), then it has been proven that no circuit representing the function with d gates exists. Using modern SAT solvers (e.g., [7]), the satisfiability of the instance (as well as the satisfying assignments) can be efficiently determined for functions with up to 6 variables.

　　For the considered adder, no solution is found for $d = 1$, $d = 2$, and $d = 3$. In contrast, for $d = 4$ a satisfying assignment is obtained leading to the circuit shown in Figure 4.3. Thus, the minimal realization (with respect to gate count) consists of four gates. The resulting circuit has quantum cost of 12^2.

4.5   HEURISTIC SYNTHESIS

In the past, several heuristic approaches for synthesizing reversible functions have been introduced [30, 21, 14, 17, 18, 12, 29, 28]. In this section, we describe the synthesis of the adder using the approach of [21] (also known as the *MMD approach*). The basic idea is to traverse each line of the truth table and adding gates to the circuit until the output values match the input values (i.e., until the identity of both is achieved). Gates are thereby chosen so that they don't alter already considered lines. Furthermore, gates are added starting at the output side of the circuit (this is because *output* values are transformed until the identity is achieved).

In the following, we describe the respective steps of the approach using Table 4.4. The first column denotes the respective line numbers, while the second and third column give the function specification of the adder (taken from Table 4.3). For brevity, the inputs 0, c_{in}, x, y and outputs c_{out}, sum, g_1, g_2 are denoted by a, b, c, d, respectively. The remaining columns provide the transformed output values for the respective steps.

line (i)	input abcd	output abcd	1st step abcd	2nd step abcd	3rd step abcd	4th step abcd	5th step abcd	6th step abcd
0	0000	0000	0000	0000	0000	0000	0000	0000
1	0001	0111	0101	0001	0001	0001	0001	0001
2	0010	0110	0110	0110	0010	0010	0010	0010
3	0011	1001	1011	1111	1011	0011	0011	0011
4	0100	0100	0100	0100	0100	0100	0100	0100
5	0101	1011	1001	1101	1101	1101	0101	0101
6	0110	1010	1010	1010	1110	1110	1110	0110
7	0111	1101	1111	1011	1111	0111	1111	0111
8	1000	1000	1000	1000	1000	1000	1000	1000
9	1001	1111	1101	1001	1001	1001	1001	1001
10	1010	1110	1110	1110	1010	1010	1010	1010
11	1011	0001	0011	0111	0011	1011	1011	1011
12	1100	1100	1100	1100	1100	1100	1100	1100
13	1101	0011	0001	0101	0101	0101	1101	1101
14	1110	0010	0010	0010	0110	0110	0110	1110
15	1111	0101	0111	0011	0111	1111	0111	1111

Table 4.4: MMD procedure.

The algorithm starts at truth table line 0. Since for this line the input is equal to the output (both are assigned to 0000), no gate has to be added. In contrast, to match the output with the input

---

[2]Note that it is possible to synthesize circuits with larger number of gates but with lower quantum cost. Besides that, also the chosen embedding of the adder may have an effect on the quantum cost. Thus, the result from Figure 4.3 is optimal with respect to the number of gates as well as with respect to the embedding of Table 4.3.

in truth table line 1, the values for c and b must be inverted. To this end, two gates $TOF(\{d\}, c)$ (1st step) and $TOF(\{d\}, b)$ (2nd step) are added as depicted in Figure 4.4. Because of the control line d, this does not affect the previous truth table line. In line 2 and line 3, a $TOF(\{c\}, b)$ gate as well as a $TOF(\{c, d\}, a)$ gate is needed to match the values of b and a, respectively, (step 3 and 4). For the latter, two control lines are needed to keep the already traversed truth table lines unaltered. Afterwards, only two more gates $TOF(\{d, b\}, a)$ (5th step) and $TOF(\{c, b\}, a)$ (6th step) are necessary to achieve the input-output identity. The resulting circuit is shown in Figure 4.4. This circuit consists of six gates and has quantum cost of 18.

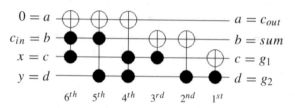

Figure 4.4: Circuit obtained by MMD synthesis.

In [21], further variations of this approach are discussed. In fact, this transformation can also be applied in the inverse direction (i.e., so that the input must match the output) and in both directions simultaneously. Furthermore, in [18] the approach has been extended by the application of templates. Templates help one to reduce the size of the resulting circuits and, thus, to achieve circuits with lower cost.

4.6   BDD-BASED SYNTHESIS FOR LARGE FUNCTIONS

The synthesis approaches described in the previous sections use a truth table description of the function to be synthesized. As a result, they are only applicable to relatively small functions. More precisely, the exact approach reaches its limits with functions containing more than 6 variables, while the MMD method is able to synthesize functions with up to 16 variables. Similar limitations can be observed if other synthesis approaches [30, 21, 14, 17, 18, 12, 29, 28] are considered. This is caused by the fact that most of them also use truth tables or apply similar strategies (namely selecting reversible gates so that the chosen function representation becomes the identity).

In this section, we describe a recently presented synthesis approach [35] that can cope with significantly larger functions. Therefore, *Binary Decision Diagrams* (BDDs) [4] are exploited. A BDD is a directed acyclic graph $G = (V, E)$ where a Shannon decomposition

$$f = \overline{x}_i f_{x_i=0} + x_i f_{x_i=1} \quad (1 \leq i \leq n)$$

is carried out in each node $v \in V$. The functions $f_{x_i=0}$ and $f_{x_i=1}$ are the *cofactors* of f. In the following, the node representing $f_{x_i=0}$ ($f_{x_i=1}$) is denoted by $low(v)$ ($high(v)$), while x_i is called the *select variable*.

It is well known that Boolean functions can be efficiently represented by BDDs [4]. Having a BDD $G = (V, E)$ representing f, a reversible circuit for this function can be derived by traversing the decision diagram and substituting each node $v \in V$ with a cascade of reversible gates. The respective cascade of gates depends on the successors of the node v. Table 4.5 provides the cascades of Toffoli gates for all possible scenarios of a BDD node.

Note that an additional (constant) line is necessary if one of the edges $low(v)$ or $high(v)$ leads to a terminal (i.e., 0 or 1) node. This is because of the irreversibility of the respective node function. As an example, consider a node v with $high(v) = 0$ (second row of Table 4.5). Without loss of generality, the first three lines of the corresponding truth table can be embedded with respect to reversibility as depicted in Table 4.7(a). However, since f is 0 in the last line, no reversible embedding for the whole function is possible (see also Section 4.3). Thus, an additional line is required to make the respective substitution reversible (see Table 4.7(b))[3].

Based on these substitutions, a method for synthesizing Boolean functions in reversible or quantum logic can be formulated: First, a BDD for function f to be synthesized is created. This can be done efficiently using state-of-the-art BDD packages (e.g., CUDD [31]). Next, the resulting BDD $G = (V, E)$ is traversed in a depth-first manner. For each node $v \in V$, cascades as depicted in Table 4.5 are added to the circuit. As an example, consider the BDD in Figure 4.5(a). Applying the substitutions given in Table 4.5, the Toffoli circuit depicted in Figure 4.5(b) results.

As a result, circuits are synthesized that realize a given function f. The capabilities of the approach are thereby bounded by the BDD size (because of the fact that each node of the BDD is substituted by a fix cascade of gates). Since well-engineered packages for BDD constructions exist (e.g., CUDD [31]), functions with more than 100 variables can be efficiently synthesized. Moreover, due to the node substitution irreversible functions are automatically embedded in a reversible description. In contrast, the resulting circuits often are sub-optimal. In particular, the number of additional lines is often significantly larger than in circuits obtained by previous synthesis approaches.

Figure 4.6 shows the resulting circuit after applying the BDD-based synthesis (with adjusted substitutions for shared nodes [4] and complement edges [3]) to the adder function. This circuit consists of ten gates and has quantum cost of 22. Furthermore, two more circuit lines are needed in comparison to the realizations shown in the previous sections.

4.7   CONCLUSIONS AND FUTURE WORK

In this chapter, we described current approaches for reversible logic synthesis. We focused on Toffoli circuit synthesis. The described approaches can also be extended to support other gates types like Fredkin or quantum gates. Besides an introduction to reversible functions and circuits, we showed how to embed irreversible functions and afterwards synthesize a Toffoli circuit using an exact and a heuristic approach. Since these approaches are limited by the truth table representation, a BDD-based synthesis approach for large functions has been described.

---

[3]For the same reason, it is also impossible to preserve the values for $low(v)$ or $high(v)$, respectively, in the substitution depicted in the first row of Table 4.5.

Table 4.5: Substitution of BDD nodes to reversible circuit.

BDD	Toffoli Circuit

Table 4.6: (Partial) Truth tables for node v with $high(v) = 0$.

x_i	$low(f)$	f	—		0	x_i	$low(f)$	f	x_i	$low(f)$
0	0	0	0		0	0	0	0	0	0
0	1	1	1		0	0	1	1	0	1
1	0	0	1		0	1	0	0	1	0
1	1	0	?		0	1	1	0	1	1
(a) w/o add. line					(b) with additional line					

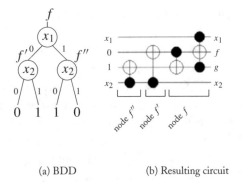

(a) BDD   (b) Resulting circuit

Figure 4.5: BDD and Toffoli circuit for $f = x_1 \oplus x_2$.

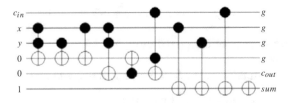

Figure 4.6: Circuit obtained by BDD approach.

Table 4.7 summarizes the benefits (+) and drawbacks (−) of the respective approaches. Exact synthesis realizes minimal circuits with respect to gate count, but is only applicable to very small functions. The MMD approach can handle somewhat larger functions, but with non-optimal results. Furthermore, both approaches are minimal with respect to the number of additional lines. In contrast, the BDD-based approach can handle significantly larger functions, but synthesizes circuits with higher gate count and more additional lines.

Using the approaches introduced in this chapter as a basis, several open questions can be addressed in future work. Even if the BDD-based approach allows efficient synthesis of functions with more than 100 variables, the resulting circuits are sub-optimal. Thus, how to further optimize

Table 4.7: Comparison of synthesis approaches.

	Circuit Size		Scalability
	Gates	Lines	
Exact approach	++	++	- -
MMD approach	o	++	o
BDD approach	-	- -	++

large reversible circuits (in particular with respect to the number of additional lines) is an important problem. Existing approaches [18, 40, 8] provide a good starting point, but currently are applicable to small circuits only.

Besides that, the ability to synthesize large (and therewith complex) circuits will raise the question if the synthesized design shows the expected behavior. Thus, verification (and afterwards debugging) may become important in the future. First approaches for verifying [33, 34, 38] and even debugging [37] reversible and quantum circuits exist, but must be further improved to really become applicable in a future design flow. Finally, the problem of testing is going to become important (first work can be found in [24, 26, 27]). Already today, small physical realizations of reversible (and quantum circuits) are available (see Section 1). For the future, larger implementations can be expected requiring efficient methods to detect production errors.

4.8   EXERCISES

4.1.   How many reversible functions over n variables exist?

4.2.   Consider the reversible circuit depicted in Figure 4.7.

　　a Specify the truth table of the function that is realized by this circuit.

　　b Calculate the quantum cost of this circuit.

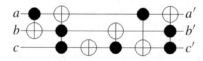

Figure 4.7: Exercise 4.2.

4.3.   Consider the *AND* function specified in Table 4.8.

　　a Why is this function irreversible?

　　b Embed this function in a reversible function.

Table 4.8: Exercise 4.3.		
a	b	o
0	0	0
0	1	0
1	0	0
1	1	1

Table 4.9: Exercise 4.4.					
a	b	c	a'	b'	c'
0	0	0	0	0	1
0	0	1	0	0	0
0	1	0	0	1	1
0	1	1	0	1	0
1	0	0	1	0	1
1	0	1	1	1	1
1	1	0	1	0	0
1	1	1	1	1	0

4.4. Consider the function specified in Table 4.9. Synthesize a circuit realizing this function using the MMD approach introduced in Section 4.5.

4.5. Again, consider the *AND* function. Synthesize a circuit realizing this function using the BDD-based approach introduced in Section 4.6.

REFERENCES

[1] A. Barenco, C. H. Bennett, R. Cleve, D. P. DiVinchenzo, N. Margolus, P. Shor, T. Sleator, J. A. Smolin, and H. Weinfurter, "Elementary gates for quantum computation," *The American Physical Society*, 52:3457–3467, 1995. 4.2

[2] C. H. Bennett, "Logical reversibility of computation," *IBM J. Res. Dev*, 17(6):525–532, 1973. 4.1

[3] K. Brace, R. Rudell, and R. Bryant, "Efficient implementation of a BDD package," In *Design Automation Conf.*, pages 40–45, 1990. DOI: 10.1145/123186.123222 4.6

[4] R. Bryant, "Graph-based algorithms for Boolean function manipulation," *IEEE Trans. on Comp.*, 35(8):677–691, 1986. DOI: 10.1109/TC.1986.1676819 4.6, 4.6

[5] R. Cuykendall and D. R. Andersen, "Reversible optical computing circuits," *Optics Letters*, 12(7):542–544, 1987. DOI: 10.1364/OL.12.000542 4.1

[6] B. Desoete and A. de Vos, "A reversible carry-look-ahead adder using control gates," *INTEGRATION, the VLSI Jour.*, 33(1-2):89–104, 2002. DOI: 10.1016/S0167-9260(02)00051-2 4.1

[7] N. Eén and N. Sörensson, "An extensible SAT solver," In *SAT 2003*, volume 2919 of *LNCS*, pages 502–518, 2004. DOI: 10.1007/b95238 4.4, 4.4

[8] D. Y. Feinstein, M. A. Thornton, and D. M. Miller, "Partially redundant logic detection using symbolic equivalence checking in reversible and irreversible logic circuits," In *Design, Automation and Test in Europe*, pages 1378–1381, 2008. DOI: 10.1145/1403375.1403707 4.7

[9] E. F. Fredkin and T. Toffoli, "Conservative logic," *International Journal of Theoretical Physics*, 21(3/4):219–253, 1982. DOI: 10.1007/BF01857727 4.2

[10] D. Große, R. Wille, G. W. Dueck, and R. Drechsler, "Exact synthesis of elementary quantum gate circuits for reversible functions with don't cares," In *Int'l Symp. on Multi-Valued Logic*, pages 214–219, 2008. DOI: 10.1109/ISMVL.2008.42 4.4

[11] D. Große, R. Wille, G. W. Dueck, and R. Drechsler, "Exact multiple control Toffoli network synthesis with SAT techniques," *IEEE Trans. on CAD*, 28(5):703–715, 2009. DOI: 10.1109/TCAD.2009.2017215 4.4

[12] P. Gupta, A. Agrawal, and N. Jha, "An algorithm for synthesis of reversible logic circuits," *IEEE Trans. on CAD*, 25(11):2317–2330, 2006. DOI: 10.1109/TCAD.2006.871622 4.5, 4.6

[13] W. Hung, X. Song, G. Yang, J. Yang, and M. Perkowski, "Optimal synthesis of multiple output Boolean functions using a set of quantum gates by symbolic reachability analysis," *IEEE Trans. on CAD*, 25(9):1652–1663, 2006. DOI: 10.1109/TCAD.2005.858352 4.4

[14] P. Kerntopf, "A new heuristic algorithm for reversible logic synthesis," In *Design Automation Conf.*, pages 834–837, 2004. DOI: 10.1145/996566.996789 4.5, 4.6

[15] R. Landauer, "Irreversibility and heat generation in the computing process," *IBM J. Res. Dev.*, 5:183, 1961. 4.1

[16] D. Maslov and G. W. Dueck, "Improved quantum cost for n-bit Toffoli gates," *IEE ELECTRONICS LETTERS*, 39:1790, 2004. DOI: 10.1049/el:20031202 4.2

[17] D. Maslov and G. W. Dueck, "Reversible cascades with minimal garbage," *IEEE Trans. on CAD*, 23(11):1497–1509, 2004. DOI: 10.1109/TCAD.2004.836735 4.3, 4.5, 4.6

[18] D. Maslov, G. W. Dueck, and D. M. Miller, "Toffoli network synthesis with templates," *IEEE Trans. on CAD*, 24(6):807–817, 2005. DOI: 10.1109/TCAD.2005.847911 4.5, 4.5, 4.6, 4.7

[19] D. Maslov, C. Young, G. W. Dueck, and D. M. Miller, "Quantum circuit simplification using templates," In *Design, Automation and Test in Europe*, pages 1208–1213, 2005. DOI: 10.1109/DATE.2005.249 4.2

[20] R. C. Merkle, "Reversible electronic logic using switches," *Nanotechnology*, 4:21–40, 1993. DOI: 10.1088/0957-4484/4/1/002 4.1

[21] D. M. Miller, D. Maslov, and G. W. Dueck, "A transformation based algorithm for reversible logic synthesis," In *Design Automation Conf.*, pages 318–323, 2003. DOI: 10.1145/775832.775915 4.5, 4.5, 4.6

[22] D. M. Miller, R. Wille, and G. W. Dueck, "Synthesizing reversible circuits for irreversible functions," In *Euromicro Conf. on Digital System Design*, pages 749–756, 2009. 4.3

[23] M. Nielsen and I. Chuang, *Quantum Computation and Quantum Information*. Cambridge Univ. Press, 2000. 4.1, 4.2, 4.2

[24] K. N. Patel, J. P. Hayes, and I. L. Markov, "Fault testing for reversible circuits," *IEEE Trans. on CAD*, 23(8):1220–1230, 2004. DOI: 10.1109/TCAD.2004.831576 4.7

[25] A. Peres, "Reversible logic and quantum computers," *Phys. Rev. A*, (32):3266–3276, 1985. DOI: 10.1103/PhysRevA.32.3266 4.2

[26] M. Perkowski, J. Biamonte, and M. Lukac, "Test generation and fault localization for quantum circuits," In *Int'l Symp. on Multi-Valued Logic*, pages 62–68, 2005. DOI: 10.1109/ISMVL.2005.46 4.7

[27] I. Polian, T. Fiehn, B. Becker, and J. P. Hayes, "A family of logical fault models for reversible circuits," In *Asian Test Symp.*, pages 422–427, 2005. DOI: 10.1109/ATS.2005.9 4.7

[28] M. Saeedi, M. Sedighi, and M. S. Zamani, "A novel synthesis algorithm for reversible circuits," In *Int'l Conf. on CAD*, pages 65–68, 2007. DOI: 10.1109/ICCAD.2007.4397245 4.5, 4.6

[29] V. V. Shende, S. S. Bullock, and I. L. Markov, "Synthesis of quantum-logic circuits," *IEEE Trans. on CAD*, 25(6):1000–1010, 2006. DOI: 10.1145/1120725.1120847 4.5, 4.6

[30] V. V. Shende, A. K. Prasad, I. L. Markov, and J. P. Hayes, "Synthesis of reversible logic circuits," *IEEE Trans. on CAD*, 22(6):710–722, 2003. DOI: 10.1145/774572.774625 4.4, 4.5, 4.6

[31] F. Somenzi, *CUDD: CU Decision Diagram Package Release 2.3.1*. University of Colorado at Boulder, 2001. 4.6, 4.6

[32] T. Toffoli, "Reversible computing," In W. de Bakker and J. van Leeuwen, editors, *Automata, Languages and Programming*, page 632. Springer, 1980. Technical Memo MIT/LCS/TM-151, MIT Lab. for Comput. Sci. 4.1, 4.2

[33] G. F. Viamontes, I. L. Markov, and J. P. Hayes, "Checking equivalence of quantum circuits and states," In *Int'l Conf. on CAD*, pages 69–74, 2007. DOI: 10.1109/ICCAD.2007.4397246 4.7

[34] S.-A. Wang, C.-Y. Lu, I.-M. Tsai, and S.-Y. Kuo, "An XQDD-based verification method for quantum circuits," *IEICE Transactions*, 91-A(2):584–594, 2008. DOI: 10.1093/ietfec/e91-a.2.584 4.7

[35] R. Wille and R. Drechsler, "BDD-based synthesis of reversible logic for large functions," In *Design Automation Conf.*, pages 270–275, 2009. DOI: 10.1145/1629911.1629984 4.6

[36] R. Wille, D. Große, G. W. Dueck, and R. Drechsler, "Reversible logic synthesis with output permutation," In *VLSI Design*, pages 189–194, 2009. DOI: 10.1109/VLSI.Design.2009.40 4.3

[37] R. Wille, D. Große, S. Frehse, G. W. Dueck, and R. Drechsler, "Debugging of Toffoli networks," In *Design, Automation and Test in Europe*, pages 1284–1289, 2009. 4.7

[38] R. Wille, D. Große, D. M. Miller, and R. Drechsler, "Equivalence checking of reversible circuits," In *Int'l Symp. on Multi-Valued Logic*, pages 324–330, 2009. DOI: 10.1109/ISMVL.2009.19 4.7

[39] V. V. Zhirnov, R. K. Cavin, J. A. Hutchby, and G. I. Bourianoff, "Limits to binary logic switch scaling – a gedanken model," *Proc. of the IEEE*, 91(11):1934–1939, 2003. DOI: 10.1109/JPROC.2003.818324 4.1

[40] J. Zhong and J. Muzio, "Using crosspoint faults in simplifying toffoli networks," In *IEEE North-East Workshop on Circuits and Systems*, pages 129–132, 2006. DOI: 10.1109/NEWCAS.2006.250942 4.7

CHAPTER 5

Data Mining Using Binary Decision Diagrams

Shin-ichi Minato

CHAPTER SUMMARY

Binary Decision Diagrams (BDDs) are an efficient data structure for representing Boolean functions. Much research has been done on BDD manipulation in VLSI logic design since 1990's. We found that BDD-based techniques can also be applied effectively to data mining and knowledge discovery. Especially, Zero-suppressed BDDs (ZDDs) are suitable for handling sets of sparse combinations that often appear in many practical database analyses. In this chapter, we show recent activities in BDD-based knowledge discovery techniques. First, we describe the "LCM over ZDDs" algorithm, which achieves very fast frequent itemset mining. Then, we present some useful post-processing methods for analyzing the results of the frequent itemset mining.

5.1 INTRODUCTION

Manipulation of large-scale combinatorial data is one of the fundamental requirements for data mining and knowledge discovery. In particular, frequent itemset analysis is important in many tasks that try to find interesting patterns from web documents and databases, such as association rules, correlations, sequences, episodes, classifiers, and clusters. Since the introduction by Agrawal et al. [1], the frequent itemset and association rule analysis have received much attention from many researchers. A number of papers have been published about new algorithms or improvements for solving such mining problems [4, 7, 21].

On the other hand, Binary Decision Diagrams (BDDs) [2] are an efficient data structure for representing Boolean functions. Since the 1990's, wide use of BDDs has occurred in VLSI logic design. Recently, we found that BDD-based techniques can also be applied effectively to problems in data mining and knowledge discovery. Especially, Zero-suppressed BDDs (ZDDs) [14] are suitable for handling sets of sparse combinations that often appear in many practical database analyses.

In this chapter, we show recent activities in ZDD-based knowledge discovery techniques. First, we describe a basic method for representing itemsets of transaction databases using ZDDs. Then, we show the "LCM over ZDDs" algorithm, which is a recent approach to very fast frequent itemset mining. This algorithm is based on the *LCM* (*Linear time Closed itemset Miner*) [20], one of the most efficient state-of-the-art techniques for itemset mining. The LCM over ZDDs algorithm can directly generate compact output data structures in main memory to be efficiently post-processed by using ZDD-based algebraic operations. Finally, we present several methods of using LCM over ZDDs and post-processing for finding distinctive itemsets from the databases.

5.2 BDD-BASED DATABASE REPRESENTATION

5.2.1 TRANSACTION DATABASES AND ITEMSET MINING

Let $\mathcal{E} = \{1, 2, \ldots, n\}$ be an ordered set of *items*[1]. A (*transaction*) *database* on $\mathcal{E}$ is a list $D = (T_1, T_2, \ldots, T_m)$ where $T_i \subseteq \mathcal{E}$ for $1 \leq i \leq m$. Each T_i in D is called a *transaction* (or *tuple*). We denote the sum of cardinalities of all T_i, by $||D|| = \Sigma_{1 \leq i \leq m} |T_i|$. $||D||$ is called the size of database D. A set $P \subseteq \mathcal{E}$ is called an *itemset*. The maximum (highest ordered) item in an itemset P is called the *tail* of P, and is denoted by $tail(P)$.

Now, we consider a particular database. For a given itemset P, a transaction T_i such that $P \subseteq T_i$ is an *occurrence* of P. The number of all occurrences of P in the database (i.e., the number of times such that $P \subseteq T_i$) is called the *frequency* (or *support*) of P, and is denoted by $frq(P)$. In particular, for an item e, $frq(\{e\})$ is the frequency of e. That is, $frq(\{e\})$ is the number of transactions in which element e appears, $0 \leq frq(\{e\}) \leq m$. For a given constant θ ($1 \leq \theta \leq m$), called the *minimum support*, itemset P is *frequent* if $frq(P) \geq \theta$. Notice that the null itemset $\{\}$ is always frequent for any θ. The problem of frequent itemset mining is to enumerate all frequent itemsets for a given database D and minimum support θ.

Figure 5.1 shows a small example of frequent itemset mining. Here we use the letters $a, b, c, \ldots$ instead of $1, 2, 3, \ldots$ to represent items, i.e., $\mathcal{E} = \{a, b, c\}$, and a set of itemsets $\{\{1, 2\}, \{2, 3\}, \{1\}, \{2\}, \{3\}, \{\}\}$ is now written as $\{ab, bc, a, b, c, \lambda\}$. (Here λ means null itemset.) In this example, the database D has 11 transactions in total. The itemset b occurs 10 times out of 11. The itemsets a, c, and ac occur 8 times, bc occurs 7 times, ac and abc occur 5 times. Thus, if we specify $\theta = 7$, $\{ab, bc, a, b, c, \lambda\}$ is obtained as the set of frequent itemsets. We can observe that when the minimum support θ is smaller, more itemsets become frequent. If $\theta = 1$, any subset of transactions can be frequent, so the number of frequent itemsets can grow exponentially with the maximal cardinality of a transaction. Usually, we first specify a sufficiently large number for θ, and if too few frequent itemsets are generated, we may slightly decrease θ until a desired number of itemsets become frequent. We then look for some interesting or unexpected item combinations from the results.

Frequent itemset mining is one of the fundamental problems in data mining and knowledge discovery. Many kinds of real-life problems can be mapped into this problem. Here are some examples.

---

[1]In principle, $\mathcal{E}$ does not need to be ordered, but many algorithms of itemset mining assume the ordering of items, for some reasons of implementation.

- In a supermarket, a number of items are displayed, and each customer purchases a set of items in his/her basket. Thus, a transaction database stores the purchase list of a large number of customers. In this case, a frequent itemset indicates a combination of closely related items, for example, beer and snacks are often purchased together. Such analysis is called *basket analysis*.

- Frequent itemset mining can also be used for analyzing a traffic accident database. In this database, each item corresponds to a property of the accidents, for example, the make of a car, daytime or night, weather, shape of the crossroads, absence or presence of a pedestrian, etc. Then, a frequent itemset shows a dangerous pattern which may cause an accident.

- We can also analyze the relationships of internet web pages by frequent itemset mining. Each item corresponds to a web page, and an itemset corresponds to a set of linked pages written in a web page. Then, a frequent itemset shows a set of strongly related web pages.

As shown above, frequent itemset mining has many applications in social, biological, medical, and digital system design areas, so it is a very important and interesting problem considered by many people. Since the pioneering work by Agrawal *et al.* [1], various algorithms have been proposed to solve the frequent itemset mining problem [4, 21].

5.2.2   BDDS AND ZDDS

Next, we briefly review decision diagrams. A *Binary Decision Diagram* (BDD) is a graph representation of a Boolean function. An example is shown in Fig. 5.2 for $F(a, b, c) = a\overline{b}c \vee \overline{a}b\overline{c}$. Given a variable ordering (in our example a, b, c), one can use Bryant's algorithm[2] to construct the BDD

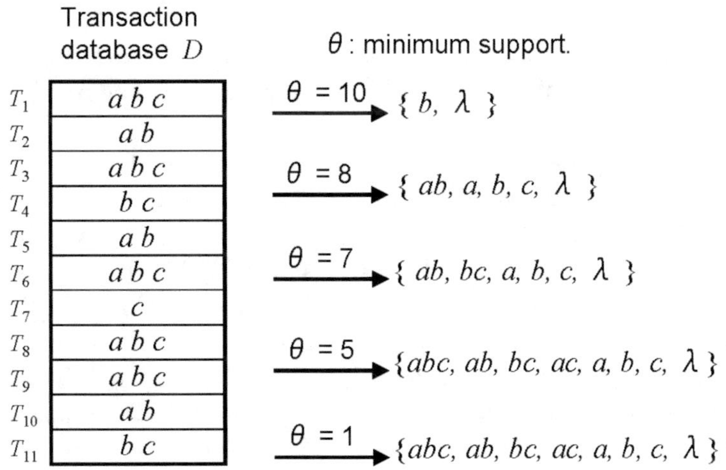

Transaction
database D θ : minimum support.

T_1	$a\ b\ c$	$\theta = 10$	$\{ b, \lambda \}$
T_2	$a\ b$		
T_3	$a\ b\ c$	$\theta = 8$	$\{ ab, a, b, c, \lambda \}$
T_4	$b\ c$		
T_5	$a\ b$	$\theta = 7$	$\{ ab, bc, a, b, c, \lambda \}$
T_6	$a\ b\ c$		
T_7	c		
T_8	$a\ b\ c$	$\theta = 5$	$\{abc, ab, bc, ac, a, b, c, \lambda \}$
T_9	$a\ b\ c$		
T_{10}	$a\ b$	$\theta = 1$	$\{abc, ab, bc, ac, a, b, c, \lambda \}$
T_{11}	$b\ c$		

Figure 5.1: Frequent itemset mining.

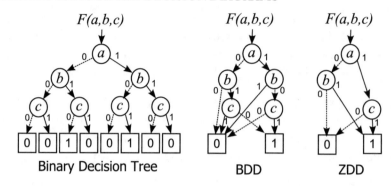

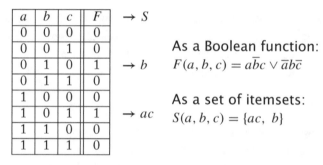

Figure 5.2: Binary Decision Tree, BDDs, and ZDDs.

a	b	c	F	→ S
0	0	0	0	
0	0	1	0	
0	1	0	1	→ b
0	1	1	0	
1	0	0	0	
1	0	1	1	→ ac
1	1	0	0	
1	1	1	0	

As a Boolean function:
$F(a, b, c) = a\bar{b}c \vee \bar{a}b\bar{c}$

As a set of itemsets:
$S(a, b, c) = \{ac,\ b\}$

Figure 5.3: Correspondence of Boolean functions and sets of combinations.

for any given Boolean function. For many Boolean functions appearing in practice, this algorithm is quite efficient, and the resulting BDDs are much more efficient representations than binary decision trees.

BDDs were originally invented to represent Boolean functions. But we can also map a set of itemsets into a Boolean space of n variables, where n is the cardinality of $\mathcal{E}$ (see Fig. 5.3). So, one could also use BDDs to represent sets of itemsets. However, one can obtain an even more efficient representation by using *Zero-suppressed BDDs* (ZDDs or ZBDDs) [14].

If there are many similar itemsets, then the subgraphs are shared resulting in a smaller representation. In addition, ZDDs have a special type of node deletion rule. In ordinary BDDs, all nodes whose two edges point to a same node are deleted, as shown in Fig. 5.4(a). Instead, ZDDs have a different rule such that all nodes whose 1-edge directly points to the 0-terminal node are deleted, as shown in Fig. 5.4(b). Because of this, the nodes of items that do not appear in any itemset are automatically deleted. This ZDD reduction rule is extremely effective if we handle a set of sparse itemsets. If the average appearance ratio of the items is 1% (i.e., $|T_i|/|\mathcal{E}| = 1\%$ on average), then ZDDs are possibly more compact than ordinary BDDs by up to 100 times.

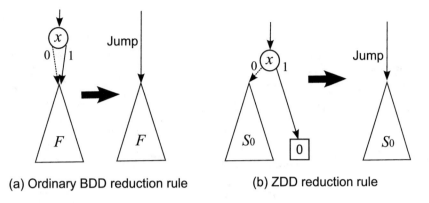

(a) Ordinary BDD reduction rule   (b) ZDD reduction rule

Figure 5.4: BDD and ZDD reduction rule.

∅	Returns empty set. (0-termial node)
{λ}	Returns singleton set of null-itemset. (1-terminal node)
S.top	Returns item-ID at root node of S.
S.offset(k)	Sub-set of itemsets not including item k.
S.onset(k)	Gets $S \setminus S$.offset(k) and then deletes item k from each itemset.
S.change(k)	Inverts existence of item k (add / delete) on each itemset.
$S_1 \cup S_2$	Returns union of the two sets.
$S_1 \cap S_2$	Returns intersection of the two sets.
$S_1 \setminus S_2$	Returns difference of the two sets. (in S_1 but not in S_2.)
S.count	Counts cardinality of S.

Figure 5.5: Primitive ZDD operations.

ZDDs have the desirable property that each path from the root node to the 1-terminal node corresponds to each itemset in the set. Namely, the number of such paths in the ZDD exactly equals to the cardinality of the set. This beautiful property indicates that, even if there are no equivalent nodes to be shared, the ZDD structure explicitly stores all itemsets as well as using an explicit linear linked list data structure. In other words, (the complexity order of) ZDD size never exceeds the explicit representation. If more nodes are shared, the ZDD is more compact than linear list.

Figure 5.5 summarizes the primitive operations of the ZDDs. In these operations, ∅, {λ}, and S.top can be obtained in a constant time. S.offset(k), S.onset(k), and S.change(k) operations require a constant time, if item k is at the root node of S. Otherwise, they consume linear time for the number of ZDD nodes located at a higher position than item k. The union, intersection, and difference operations can be performed in time that is almost linear in the size of the ZDDs. S.count is also linear to the ZDD size, and does not depend on the cardinality.

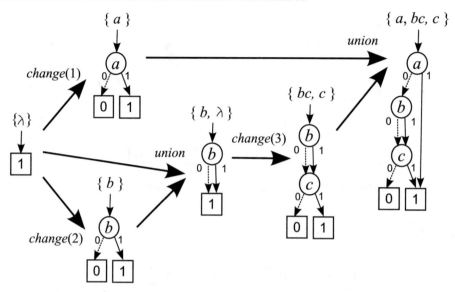

Figure 5.6: Construction of ZDDs.

Any ZDD for a set of itemsets can be generated by using a sequence of those primitive operations. Figure 5.6 shows an example of constructing a ZDD for $\{a, bc, c\}$. At first, we represent a singleton set $\{\lambda\}$ just by a 1-terminal node. Next, we generate ZDDs for $\{a\}$ and $\{b\}$ by change operations. A ZDD for $\{b, \lambda\}$ is generated by a union operation as $\{\lambda\} \cup \{b\}$. $\{bc, c\}$ can be generated by a change operation from $\{b, \lambda\}$. Finally, a union of $\{a\}$ and $\{bc, c\}$ generates $\{a, bc, c\}$. In this way, we can construct any desired ZDD by using a sequence of set operations.

Knuth [11] has a section devoted entirely to BDDs with a total of 140 pages, including 236 exercises. ZDDs are also discussed in detail using 30 pages, including 70 exercises. He re-arranged a set of primitive ZDD operations and named it "Family Algebra". His own BDD/ZDD package is available on his home page [12].

5.3   LCM OVER ZDDS FOR LARGE-SCALE ITEMSET MINING

We briefly review the "LCM over ZDDs" algorithm that efficiently generates a huge number of frequent itemsets.

5.3.1   LCM ALGORITHM

As described above, various algorithms have been proposed to solve the frequent itemset mining problem. Among those state-of-the-art algorithms, *LCM (Linear time Closed itemset Miner)* [20] by Uno et al. is an output linear time algorithm (the time complexity is linearly bounded by the output size). Their open source code[19] is considered to be one of the fastest implementations of frequent itemset mining.

In general, frequent itemset mining algorithms are classified into two categories: *apriori-like* (or *level-by-level*) algorithms [1] and *backtracking* (or *depth-first*) algorithms [7, 21]. LCM algorithms belong to the backtracking style. Backtracking algorithms are based on recursive calls. The algorithm inputs a frequent itemset P, and generates new itemsets by adding one of the unused items to P. Then, for each itemset being frequent among them, it generates recursive calls with respect to it. To avoid duplications, an iteration of the backtracking algorithms adds items with indices larger than the tail of P. Figure 5.7 shows the basic structure of the original LCM algorithm. Here, we start from $P = \{\}$ (null itemset), and we assume $tail(\{\}) = 0$.

```
LCM_Backtrack(P: itemset)
{
    Output P
    For e = n to tail(P) + 1 step −1 do
        If P ∪ {e} is frequent
            LCM_Backtrack(P ∪ {e})
}
```

```
ZDD LCMovZDD(P: itemset)
{
    ZDD F ← {λ}
    For e = n to tail(P) + 1 step −1 do
        If P ∪ {e} is frequent {
            F' ← LCMovZDD(P ∪ {e})
            F ← F ∪ F'.change(e)
        }
    Return F
}
```

Figure 5.7: Basic structure of LCM algorithm.   Figure 5.8: Basic structure of LCM over ZDDs.

LCM and most of the other itemset mining algorithms focus on just enumerating or listing the itemsets that satisfy the given conditions. It is a different matter to store and index the result of itemsets for efficient data analysis. If we want to post-process the mining results by applying various conditions or restrictions, we have to dump the frequent itemsets into sequential file storage. Even though LCM is an output linear time algorithm, time and space may be too large if the number of frequent itemsets is large.

5.3.2   LCM OVER ZDDS

Recently, Minato et al. [18] proposed a fast algorithm for generating a huge number of frequent itemsets using ZDDs. Their method, "LCM over ZDDs" is based on the LCM algorithm and it generates a compact output data using ZDDs in main memory. The result can efficiently be post-processed using algebraic ZDD operations.

The basic structure of LCM over ZDDs is described in Fig. 5.8. The original LCM (Fig. 5.7) explores all the candidate itemsets in a backtracking (or depth-first) manner. When a frequent itemset is found, it is appended to the output file. On the other hand, LCM over ZDDs constructs a ZDD that is the union of all the itemsets found in the backtracking search, and finally returns a pointer to the root node of the ZDD. Exercise 1.7 considers a trace of LCM over ZDDs algorithm. It will be helpful for understanding the action of the algorithm.

We can observe a number of common properties in LCM algorithms and ZDD manipulation. They are as follows:

- Both are based on the backtracking (depth-first) algorithm.

- All the items used in the database have a fixed variable ordering.

- In the algorithm, we choose items one by one according to the variable ordering, and then recursively call the algorithm.

- In the current LCM implementation, the variable ordering is decided at the beginning of the algorithm, and the ordering is never changed until the end of execution.

These common properties indicate that LCM and ZDDs will be a good combination. Because the original LCM algorithm contains plenty of techniques for fast access of databases, LCM over ZDDs does not change the core algorithm of LCM, and just generates a ZDD for the solutions obtained by LCM.

The performance of LCM over ZDDs is reported in [18]. Here, we summarize the results. Table 5.1 shows the specifications of the benchmark datasets, chosen from the FIMI2003 repository [5]. Table 5.2 shows the performance comparison between the original LCM and LCM over ZDDs. |ZDD| represents the number of ZDD nodes representing all the frequent itemsets. The column "LCM-count" shows the computation time of the original LCM when counting only the number of itemsets, and "LCM-dump" represents the time for listing all the itemsets into the output file (using /dev/null). "LCM over ZDDs" shows the time for generating the results of the ZDD in main memory, including the time for counting the ZDD nodes.

We can observe that LCM over ZDDs is more efficient than the original LCM-dump, even if LCM-dump uses /dev/null without actual hard disk access. The difference becomes significant when a very large number of itemsets are generated. The original LCM-dump is known to be an output linear time algorithm, but LCM over ZDDs requires a sub-linear time for the number of itemsets. The computational time of LCM over ZDDs is almost the same as executing an LCM-count. We must emphasize that LCM-count does not store the itemsets, but only counts the number of solutions. On the other hand, LCM over ZDDs generates all the solutions and stores them in main memory as a compact ZDD. This is an important difference.

5.4   APPLICATION OF LCM OVER ZDDS FOR FINDING DISTINCTIVE ITEMSETS

After generating a set of frequent itemsets, we sometimes want to filter the result according to whether a specific sub-patten P is included or not, in order to discover some useful knowledge. Those operations are called constraint itemset mining. In conventional methods, it is too time

Table 5.1: Benchmark examples for frequent itemset mining.

Database name	#Items	#Trans- actions	$\|D\|$ (size of D)
mushroom	119	8,124	186,852
BMS-WebView-1	497	59,602	149,639
BMS-WebView-2	3,340	77,512	358,278
T10I4D100K	870	100,000	1,010,228
chess	75	3,196	118,252
connect	129	67,557	2,904,951
pumsb	2,113	49,046	3,629,404

Table 5.2: Comparison of LCM over ZDDs with original LCM.

Database name	Min. support	#Frequent itemsets	LCM over ZDDs |ZDD|	Time(s)	LCM-count Time(s)	LCM-dump Time(s)
mushroom	1,000	123,287	760	0.50	0.49	0.64
	300	5,259,786	4,412	2.25	2.22	9.96
	100	66,076,586	11,584	5.06	4.87	114.21
	50	198,169,866	17,830	8.17	7.86	357.27
BMS-WebView-1	50	8,192	3,415	0.11	0.11	0.12
	40	48,544	10,755	0.18	0.18	0.22
	36	461,522	28,964	0.49	0.42	0.98
	34	4,849,466	49,377	1.30	1.07	8.58
	32	1,531,980,298	71,574	31.90	29.73	3,843.06
BMS-WebView-2	5	26,946,004	353,091	4.84	3.62	51.28
T10I4D100K	2	19,561,715	3,270,977	9.68	5.09	22.66
chess	1,000	29,442,849	53,338	197.58	197.10	248.18
connect	40,000	23,981,184	3,067	5.42	5.40	49.21
pumsb	32,000	7,733,322	5,443	60.65	60.42	75.29

(2.4GHz Core2Duo E6600 PC, 2 GB memory, SuSE Linux 10, GNU C++)

consuming to generate all frequent itemsets with a lower minimum support, so we consider only highly frequent itemsets whose size seems feasible for human beings, and then apply the filtering. This is a limitation of the conventional methods because it is difficult to find itemsets that are not very highly frequent but have a distinct property.

By using LCM over ZDDs, a huge number of itemsets can be stored and indexed compactly in main memory. In many cases, therefore, it is possible to generate all frequent itemsets with a lower minimum support and then post-process them using algebraic ZDD operations. Here, we present several post-processing operations for the sets of frequent itemsets.

5.4.1 SUBPATTERN MATCHING FOR THE FREQUENT ITEMSETS

From the frequent itemset results F, we can efficiently filter a subset S, such that each itemset in S contains a given sub-pattern (itemset) P. This is done by the ZDD operation

$S \leftarrow F$
forall $v \in P$ **do:**
$S \leftarrow S.\text{onset}(v).\text{change}(v)$
return S

Conversely, we can extract a subset of the itemsets not satisfying the given conditions. This is easily done by computing ZDD-based set operation: $F \setminus S$. Exercise 1.9 shows an example of the use of those sub-pattern filtering operations. The computation time for the sub-pattern matching is much smaller than the time for frequent itemsets mining.

The above operations are sometimes called constraint pattern mining. In conventional methods, it is too time consuming to generate all frequent itemsets before filtering. Therefore, many researchers consider direct methods of constraint pattern mining without generating all frequent itemsets. However, using the ZDD-based method, a large set of itemsets can be stored and indexed compactly in main memory. In many cases, therefore, it is possible to generate all frequent itemsets and then post-process them using algebraic ZDD operations.

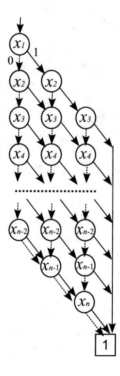

Figure 5.9: A ZDD for all combinations of less than 4 out of n items.

5.4.2   EXTRACTING LONG/SHORT ITEMSETS

Sometimes we are interested in the long/short itemsets, comprising a large/small number of items. Using ZDDs, all combinations of less than k out of n items are efficiently represented in polynomial size, bounded by $O(k \cdot n)$. For example, Fig. 5.9 shows a ZDD representing all combinations of less than 4 out of n items.

Such a ZDD represents a length constraint on itemsets. We then apply an intersection (or difference) operation to the frequent itemsets that meet the length constraint of the ZDD. In this way, we can easily extract a set of long/short frequent itemsets.

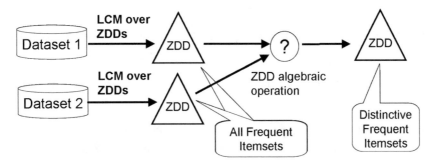

Figure 5.10: Comparison of two frequent itemsets.

5.4.3   COMPARISON BETWEEN TWO SETS OF FREQUENT ITEMSETS

Our ZDD manipulation environment can efficiently store more than one set of results of frequent itemset mining. Therefore, we can compare two sets of frequent itemsets generated under different conditions. As shown in Fig. 5.10, our ZDD-based method can store and index two sets of frequent itemsets for the datasets from different times, and easily compute the intersection, union, and difference between the frequent itemsets. Typical usage of this method is to compare a current database with a past one, in order to discover changes.

When we use a conventional representation of frequent itemsets, we could generate and store a limited number of frequent itemsets controlled by increasing the support parameter, and thus, we could only detect the changes of very highly frequent itemsets. On the other hand, ZDD-based representation enables us to generate and store a huge number of itemsets which are relatively less frequent, therefore, we can explore many more interesting patterns.

5.4.4   CALCULATING STATISTICAL DATA

After generating a ZDD for a set of itemsets, we can quickly count the number of itemsets by using a primitive ZDD operation. The computation time is linearly bounded by the ZDD size, and not by the itemset count. We can also efficiently calculate other statistical measures, such as *Support* and *Confidence*, which are often used in probabilistic analysis and machine learning.

5.5 CONCLUSION

In this chapter, we described our recent research activities on BDD-based data processing for data mining and knowledge discovery. BDDs and ZDDs provide automatic compressed graph representations for a huge number of itemsets. The compressed representation can be post-processed and analyzed efficiently by using various set operations without decompression.

Our future work includes the following:

- Considering a series of itemset operations to extract interesting information from practical databases.

- Developing a good user interface for queries to the database, accommodating the kind of information desired by the user.

- Applying the method not only for transaction databases but also for sequential or streaming databases.

We expect that the ZDD-based data processing techniques will become more popular in the field of knowledge discovery, and now we can see some encouraging signs for applying to various real-life problems.

5.6 EXERCISES

5.1. Calculate $||D||$ for the example of Fig. 5.1.

5.2. Let $P = \{2, 4, 5\}$. What is $tail(P)$?

5.3. For the example of Fig. 5.1, enumerate all the occurrences of itemset bc.

5.4. For the example of Fig. 5.1, show the frequent itemsets for the minimum support $\theta = 6$.

5.5. For the example of Fig. 5.1, describe ZDDs for the frequent itemsets, with minimum support $\theta = 10, 8, 7$, and 5.

5.6. Also, for the example of Fig. 5.1, describe (ordinary) BDDs for the frequent itemsets, with minimum support $\theta = 10$ and 5, and then compare the BDDs and ZDDs.

5.7. For the example of Fig. 5.1, trace the algorithms of LCM over ZDDs shown in Fig. 5.8, with $\theta = 7$.

5.8. For each benchmark dataset shown in Table 5.1, compute the average appearance ratio of the items (i.e., average of $|T_i|/|\mathcal{E}|$).

5.9. From the answer of 5.4, filter a set of itemsets S such that S includes b but does not include c.

5.10. As well as Fig. 5.9, describe a ZDD representing all combinations of exactly 3 out of *n* items.

5.11. For the example of Fig. 5.1, consider the way to generate a ZDD representing all the itemsets whose frequencies are more than 6 and less than 10.

REFERENCES

[1] R. Agrawal, T. Imielinski, and A. N. Swami, "Mining association rules between sets of items in large databases," In P. Buneman and S. Jajodia, editors, *Proc. of the 1993 ACM SIGMOD International Conference on Management of Data, Vol. 22(2) of SIGMOD Record*, pages 207–216, 1993. DOI: 10.1145/170035 5.1, 5.2.1, 5.3.1

[2] R. E. Bryant, "Graph-based algorithms for Boolean function manipulation," *IEEE Transactions on Computers*, C-35(8):677–691, 1986. DOI: 10.1109/TC.1986.1676819 5.1, 5.2.2

[3] R. E. Bryant and Y.-A. Chen, "Verification of arithmetic circuits with binary moment diagrams," In *Proc. of 32th ACM/IEEE Design Automation Conference (DAC-95)*, pages 535–541, 1995. DOI: 10.1145/217474.217583

[4] B. Goethals, "Survey on frequent pattern mining," 2003.
http://www.cs.helsinki.fi/u/goethals/publications/survey.ps 5.1, 5.2.1

[5] B. Goethals and M. J. Zaki, "Frequent itemset mining dataset repository," 2003. *Frequent Itemset Mining Implementations (FIMI'03)*,
http://fimi.cs.helsinki.fi/data/ 5.3.2

[6] J. Han, J. Pei, and Y. Yin, "Mining frequent patterns without candidate generation," In *In Proc of the 2000 ACM SIGMOD international conference on Management of data*, pages 1–12, 2000. DOI: 10.1145/342009.335372

[7] J. Han, J. Pei, Y. Yin, and R. Mao, "Mining frequent patterns without candidate generation: a frequent-pattern tree approach," *Data Mining and Knowledge Discovery*, 8(1):53–87, 2004. DOI: 10.1023/B:DAMI.0000005258.31418.83 5.1, 5.3.1

[8] M. Ishihata, Y. Kameya, T. Sato, and S. Minato, "Propositionalizing the em algorithm by BDDs," In *In Proc of 18th International Conference on Inductive Logic Programming (ILP 2008)*, 9 2008.

[9] K.-F. Jea and M.-Y. Chang, "Discovering frequent itemsets by support approximation and itemset clustering," *Data & Knowledge Engineering*, 65(1):90–107, 2008. DOI: 10.1016/j.datak.2007.10.003

[10] K.-F. Jea, M.-Y. Chang, and K.-C. Lin, "An efficient and flexible algorithm for on-line mining of large itemsets," *Information Processing Letters*, 92(6):311–316, 2004. DOI: 10.1016/j.ipl.2004.09.008

[11] D. E. Knuth, *The Art of Computer Programming: Bitwise Tricks & Techniques; Binary Decision Diagrams*, volume 4, fascicle 1. Addison-Wesley, 2009. 5.2.2

[12] D. E. Knuth, *Don Knuth's Home Page*. http://www-cs-faculty.stanford.edu/~knuth/ 5.2.2

[13] E. Loekito and J. Bailey, "Are zero-suppressed binary decision diagrams good for mining frequent patterns in high dimensional datasets?" In *Proc. of the Sixth Australasian Data Mining Conference (AusDM 2007)*, pages 135–146, 2007.

[14] S. Minato, "Zero-suppressed BDDs for set manipulation in combinatorial problems," In *Proc. of 30th ACM/IEEE Design Automation Conference*, pages 272–277, 1993. DOI: 10.1145/157485.164890 5.1, 5.2.2

[15] S. Minato and H. Arimura, "Efficient combinatorial item set analysis based on zero-suppressed BDDs," In *Proc. IEEE/IEICE/IPSJ International Workshop on Challenges in Web Information Retrieval and Integration (WIRI-2005)*, pages 3–10, 4 2005.

[16] S. Minato, K. Satoh, and T. Sato, "Compiling bayesian networks by symbolic probability calculation based on Zero-suppressed BDDs," In *Proc. of 20th International Joint Conference of Artificial Intelligence (IJCAI-2007)*, pages 2550–2555, 2007.

[17] S. Minato and N. Spyratos, "Combinatorial keyword query processing under a taxonomy model using Binary Decision diagrams," In *In workshop note of Perspectives of Intelligent System's Assistance (PISA-2007)*, 8 2007.

[18] S. Minato, T. Uno, and H. Arimura, "LCM over ZBDDs: Fast generation of very large-scale frequent itemsets using a compact graph-based representation," In *Proc. of 12-th Pacific-Asia Conference on Knowledge Discovery and Data Mining (PAKDD 2008), (LNAI 5012, Springer)*, pages 234–246, 5 2008. DOI: 10.1007/978-3-540-68125-0_22 5.3.2, 5.3.2

[19] T. Uno and H. Arimura, "Program codes of takeaki uno and hiroki arimura," 2007. http://research.nii.ac.jp/~uno/codes.htm 5.3.1

[20] T. Uno, Y. Uchida, T. Asai, and H. Arimura, "LCM: an efficient algorithm for enumerating frequent closed item sets," In *Proc. Workshop on Frequent Itemset Mining Implementations (FIMI'03)*, 2003. http://fimi.cs.helsinki.fi/src/ 5.1, 5.3.1

[21] M. J. Zaki, "Scalable algorithms for association mining," *IEEE Trans. Knowl. Data Eng.*, 12(2):372–390, 2000. DOI: 10.1109/69.846291 5.1, 5.2.1, 5.3.1

APPENDIX A

Solutions

A.1 CHAPTER 1

1.1 Functions f_1 and f_2 are P-equivalent, since $x_2 \leftrightarrow x_3$, then $f_1^P = x_1 x_3 \vee x_2 = f_2$. To consider the function f_3, we first simplify it as

$$f_3 = x_1 \bar{x}_3 \vee x_2 x_3 \vee \bar{x}_1 x_2 = x_1 \bar{x}_3 \vee x_2 (\bar{x}_1 \vee x_3).$$

Remember that $A \vee \overline{A} B = A \vee B$, where $A = x_1 \bar{x}_3$ and $B = x_2$. Thus, $f_3 = x_2 \vee x_1 \bar{x}_3$.

Functions f_1 and f_3 are NP-equivalent, since $x_2 \leftrightarrow \bar{x}_2$, then $f_1^N = x_1 \bar{x}_2 + x_3$, and further, if $x_2 \leftrightarrow x_3$, then $f_1^{NP} = f_3 = x_1 \bar{x}_3 \vee x_2 = f_4$. Note that functions f_2 and f_3 are N-equivalent, due to $x_2 \leftrightarrow x_3$.

1.2 There is only one function in the same N-equivalence class as the given function, $f_{n_1} = \bar{x}_1$.

Three functions are NP-equivalent to the given function: $f_{np_1} = \bar{x}_1$, $f_{np_2} = x_2$, $f_{np_3} = \bar{x}_2$.

1.3 Functions that belong to the same N-equivalence class as the given function are as follows: $f_{n_1} = x_1 \vee x_2$, $f_{n_2} = \bar{x}_1 \vee x_2$, $f_{n_3} = x_1 \vee \bar{x}_2$, $f_{n_4} = \bar{x}_1 \vee \bar{x}_2$.

The following functions are equivalent due to the permutation and negation of variables

$$\begin{aligned}
f_{np_1} &= x_1 \vee x_2 & f_{np_5} &= x_1 \vee x_3 & f_{np_9} &= x_2 \vee x_3, \\
f_{np_2} &= \bar{x}_1 \vee x_2 & f_{np_6} &= \bar{x}_1 \vee x_3 & f_{np_{10}} &= \bar{x}_2 \vee x_3, \\
f_{np_3} &= x_1 \vee \bar{x}_2 & f_{np_7} &= x_1 \vee \bar{x}_3 & f_{np_{11}} &= x_2 \vee \bar{x}_3, \\
f_{np_4} &= \bar{x}_1 \vee \bar{x}_2 & f_{np_8} &= \bar{x}_1 \vee \bar{x}_3 & f_{np_{12}} &= \bar{x}_2 \vee \bar{x}_3.
\end{aligned}$$

Functions obtained by negation of the output of the given function are:

$$\begin{aligned}
f_{npn_1} &= \bar{x}_1 \bar{x}_2 & f_{npn_5} &= \bar{x}_1 \bar{x}_3 & f_{npn_9} &= \bar{x}_2 \bar{x}_3 \\
f_{npn_2} &= x_1 \bar{x}_2 & f_{npn_6} &= x_1 \bar{x}_3 & f_{npn_{10}} &= x_2 \bar{x}_3 \\
f_{npn_3} &= \bar{x}_1 x_2 & f_{npn_7} &= \bar{x}_1 x_3 & f_{npn_{11}} &= \bar{x}_2 x_3 \\
f_{npn_4} &= x_1 x_2 & f_{npn_8} &= x_1 x_3 & f_{npn_{12}} &= x_2 x_4
\end{aligned}$$

All these functions f_{n_i}, f_{np_i}, and f_{npn_i} belong to the same NPN-class.

1.4 Function $f_2(x_1, x_2, x_3, x_4, x_5) = x_1 x_3 \vee \bar{x}_1 x_2 \vee x_4 x_5 \vee x_2 \bar{x}_5$ is P-equivalent to the given function, since $x_2 \leftrightarrow x_3$. Function $f_3(x_1, x_2, x_3, x_4, x_5) = x_1 x_3 \vee \bar{x}_1 x_2 \vee x_4 \bar{x}_5 \vee x_2 x_5$ is NP-

equivalent to the given function since $x_2 \leftrightarrow x_3$, and $x_5 \leftrightarrow \bar{x}_5$. The complement of the function f_3 is an example of an NPN-equivalent to the given function.

1.5 The function $f(x_1, x_2, x_3) = x_1 \vee x_2 \vee x_3$ is invariant to the permutation of variables. Each variable can be negated, which provides 8 equivalent functions. Negation of the output brings another 8 equivalent functions.

1.6 The truth-vectors of functions f_1 and f_2 are

$$\mathbf{F}_1 = [0, 1, 0, 1, 0, 1, 1, 1]^T,$$
$$\mathbf{F}_2 = [0, 0, 1, 1, 1, 0, 1, 1]^T.$$

There are $2^n = 2^3 = 8$ possible different FPRM expressions for each switching function of $n = 3$ variables. For the function f_1, these expressions are

$$E_{f_1,000} = x_3 \oplus x_2x_3 \oplus x_1x_2x_3, \quad (3/6)$$
$$E_{f_1,001} = 1 \oplus \bar{x}_3 \oplus x_1x_2\bar{x}_3, \quad (3/5)$$
$$E_{f_1,010} = x_3 \oplus x_1 \oplus x_1x_3 \oplus x_1\bar{x}_2 \oplus x_1\bar{x}_2x_3, \quad (5/9)$$
$$E_{f_1,011} = 1 \oplus \bar{x}_3 \oplus x_1\bar{x}_3 \oplus x_1\bar{x}_2\bar{x}_3, \quad (4/7)$$
$$E_{f_1,100} = x_3 \oplus x_2 \oplus x_2x_3 \oplus \bar{x}_1x_2 \oplus \bar{x}_1x_2x_3, \quad (5/9)$$
$$E_{f_1,101} = 1 \oplus \bar{x}_3 \oplus x_2\bar{x}_3 \oplus \bar{x}_1x_2\bar{x}_3, \quad (4/7)$$
$$E_{f_1,110} = 1 \oplus \bar{x}_2 \oplus \bar{x}_2x_3 \oplus \bar{x}_1 \oplus \bar{x}_1x_3 \oplus \bar{x}_1\bar{x}_2 \oplus \bar{x}_1\bar{x}_2x_3, \quad (7/12)$$
$$E_{f_1,111} = 1 \oplus \bar{x}_2\bar{x}_3 \oplus \bar{x}_1\bar{x}_3 \oplus \bar{x}_1\bar{x}_2\bar{x}_3. \quad (4/8)$$

The indices show the polarity, and the numbers in brackets (p/l) show the number of product terms p and literals l, respectively.

For the function f_2, the FPRM expressions are

$$E_{f_2,000} = x_2 \oplus x_1 \oplus x_1x_3 \oplus x_1x_2 \oplus x_1x_2x_3, \quad (5/9)$$
$$E_{f_2,001} = x_2 \oplus x_1\bar{x}_3 \oplus x_1x_2\bar{x}_3, \quad (3/6)$$
$$E_{f_2,010} = 1 \oplus \bar{x}_2 \oplus x_1\bar{x}_2 \oplus x_1\bar{x}_2x_3, \quad (4/7)$$
$$E_{f_2,011} = 1 \oplus \bar{x}_2 \oplus x_1\bar{x}_2\bar{x}_3, \quad (3/5)$$
$$E_{f_2,100} = 1 \oplus x_3 \oplus x_2x_3 \oplus \bar{x}_1 \oplus \bar{x}_1x_3 \oplus \bar{x}_1x_2 \oplus \bar{x}_1x_2x_3, \quad (7/12)$$
$$E_{f_2,101} = \bar{x}_3 \oplus x_2 \oplus x_2\bar{x}_3 \oplus \bar{x}_1\bar{x}_3 \oplus \bar{x}_1x_2\bar{x}_3, \quad (5/9)$$
$$E_{f_2,110} = 1 \oplus \bar{x}_2x_3 \oplus \bar{x}_1\bar{x}_2 \oplus \bar{x}_1\bar{x}_2x_3, \quad (4/8)$$
$$E_{f_2,111} = 1 \oplus \bar{x}_2 \oplus \bar{x}_2\bar{x}_3 \oplus \bar{x}_1\bar{x}_2\bar{x}_3. \quad (4/7)$$

These expressions correspond to $2^n = 2^3 = 8$ Reed-Muller spectra.

For function f_1 these spectra are

$$
\begin{aligned}
\mathbf{E}_{f_1,000} &= [0, 1, 0, 0, 0, 0, 1, 1]^T, \\
\mathbf{E}_{f_1,001} &= [1, 1, 0, 0, 0, 0, 0, 1]^T, \\
\mathbf{E}_{f_1,010} &= [0, 1, 0, 0, 1, 1, 1, 1]^T, \\
\mathbf{E}_{f_1,011} &= [1, 1, 0, 0, 0, 1, 0, 1]^T, \\
\mathbf{E}_{f_1,100} &= [0, 1, 1, 1, 0, 0, 1, 1]^T, \\
\mathbf{E}_{f_1,101} &= [1, 1, 0, 1, 0, 0, 0, 1]^T, \\
\mathbf{E}_{f_1,110} &= [1, 0, 1, 1, 1, 1, 1, 1]^T, \\
\mathbf{E}_{f_1,111} &= [1, 0, 0, 1, 0, 1, 0, 1]^T.
\end{aligned}
$$

For function f_2 all possible FPRM spectra are

$$
\begin{aligned}
\mathbf{E}_{f_2,000} &= [0, 0, 1, 0, 1, 1, 1, 1]^T, \\
\mathbf{E}_{f_2,001} &= [0, 0, 1, 0, 0, 1, 0, 1]^T, \\
\mathbf{E}_{f_2,010} &= [1, 0, 1, 0, 0, 0, 1, 1]^T, \\
\mathbf{E}_{f_2,111} &= [1, 0, 1, 0, 0, 0, 0, 1]^T, \\
\mathbf{E}_{f_2,100} &= [1, 1, 0, 1, 1, 1, 1, 1]^T, \\
\mathbf{E}_{f_2,101} &= [0, 1, 1, 1, 0, 1, 0, 1]^T, \\
\mathbf{E}_{f_2,110} &= [1, 0, 0, 1, 0, 0, 1, 1]^T, \\
\mathbf{E}_{f_2,111} &= [1, 0, 1, 1, 0, 0, 0, 1]^T.
\end{aligned}
$$

The vector W_p represents the number of product terms in each FPRM expression of the given function. Likewise, the vector W_l represents the number of literals in each FPRM expression. Therefore, for f_1, the Walsh spectra in descending order are $W_{p_1} = [7, 5, 5, 4, 4, 4, 3, 3]^T$ and $W_{l_1} = [12, 9, 9, 8, 7, 7, 6, 5]^T$. As expected, for f_2 they are $W_{p_2} = [7, 5, 5, 4, 4, 4, 3, 3]^T$ and $W_{l_2} = [12, 9, 9, 8, 7, 7, 6, 5]^T$.

1.7 The majority function for $n = 3$, $f_m = x_1x_2 \vee x_1x_3 \vee x_2x_3$ is a self-dual function, since $x_1x_2 \vee x_1x_3 \vee x_2x_3 = \overline{(\bar{x}_1\bar{x}_2 \vee \bar{x}_1\bar{x}_3 \vee \bar{x}_2\bar{x}_3)}$.

1.8 The majority function for $n = 3$ is a self-dual function. The SD-class defined by this function therefore includes a number of $n = 2$ functions. We can express the majority function for $n = 3$ in the following way:

$$
\begin{aligned}
f_m &= x_1x_2 \vee x_1x_3 \vee x_2x_3 \\
&= x_1(x_2 \vee x_3) \vee x_2x_3 \\
&= (x_2 \vee x_3)x_1 \vee x_2x_3\bar{x}_1 \\
&= (x_2 \vee x_3)x_1 \vee \overline{(\bar{x}_2 \vee \bar{x}_3)}\bar{x}_1.
\end{aligned}
$$

Therefore, function $f = x_2 \vee x_3$ belongs to the SD-class defined by f_m.

Furthermore, any function obtained by permutation of variables in $x_1(x_2 \lor x_3) \lor x_2 x_3 \bar{x}_1$ belongs to the same SD-class. For example, after the exchange of variables $x_1 \leftrightarrow x_2$, $f_m = (x_1 \lor x_3)x_2 \lor x_1 x_3 \bar{x}_2 = (x_1 \lor x_3)x_2 \lor (\bar{x}_1 \lor \bar{x}_3)\bar{x}_2$. The function $f = x_1 \lor x_3$ also belongs to the SD-class defined by f_m.

1.9 With the introduction of non-self-dualized variable x_4 the function $f_1 = x_1 \lor x_2 \lor x_3$ leads to

$$
\begin{aligned}
f_1^{SD} &= (x_1 \lor x_2 \lor x_3)x_4 \lor \overline{(\bar{x}_1 \lor \bar{x}_2 \lor \bar{x}_3)}\bar{x}_4 \\
&= (x_1 \lor x_2 \lor x_3)x_4 \lor x_1 x_2 x_3 \bar{x}_4.
\end{aligned}
$$

In the same way $f_2 = x_2(x_1 \lor x_3)$ produces

$$
\begin{aligned}
f_2^{SD} &= x_2(x_1 \lor x_3)x_4 \lor \overline{\bar{x}_2(\bar{x}_1 \lor \bar{x}_3)}\bar{x}_4 \\
&= x_1 x_2 x_4 \lor x_2 x_3 x_4 \lor (x_2 \lor x_1 x_3)\bar{x}_4 \\
&= x_1 x_2 x_4 \lor x_2 x_3 x_4 \lor x_2 \bar{x}_4 \lor x_1 x_3 \bar{x}_4 \\
&= (x_1 \lor x_3 \lor \bar{x}_4)x_2 \lor x_1 x_3 \bar{x}_4.
\end{aligned}
$$

After permutations $\bar{x}_4 \leftrightarrow x_4$ and $x_2 \leftrightarrow x_4$ it follows $f_1^{SD} = f_2^{SD}$, therefore, functions f_1 and f_2 belong to the same SD-class.

1.10 An LP-equivalent of the function $f_1 = x_1 \lor \bar{x}_2$ can be obtained by multiplying the truth-vector of this function with a Kronecker product of any two nonsingular matrices over $GF(2)$.
For example, let

$$
\mathbf{G} = \begin{bmatrix} 1 & 1 \\ 0 & 1 \end{bmatrix} \otimes \begin{bmatrix} 1 & 1 \\ 0 & 1 \end{bmatrix} = \begin{bmatrix} 1 & 1 & 1 & 1 \\ 0 & 1 & 0 & 1 \\ 0 & 0 & 1 & 1 \\ 0 & 0 & 0 & 1 \end{bmatrix}, \tag{A.1}
$$

and truth-vector of $f_1(x_1, x_2)$ is $\mathbf{F}_1 = [1011]^T$.
The truth-vector of f_2 is

$$
\mathbf{F}_2 = \mathbf{G}\mathbf{F}_1 = \begin{bmatrix} 1 & 1 & 1 & 1 \\ 0 & 1 & 0 & 1 \\ 0 & 0 & 1 & 1 \\ 0 & 0 & 0 & 1 \end{bmatrix} \begin{bmatrix} 1 \\ 0 \\ 1 \\ 1 \end{bmatrix} = \begin{bmatrix} 1 \\ 1 \\ 0 \\ 1 \end{bmatrix}. \tag{A.2}
$$

It follows that $f_2 = \bar{x}_1 \lor x_2$ belongs to the same LP-class as $f_1 = x_1 \lor \bar{x}_2$.

1.11 The truth-vector of $f_1 = x_1 \vee x_2$ is $\mathbf{F}_1 = [0, 1, 1, 1]^T$. The truth-vector for $f_2 = \overline{x}_1 \oplus x_1 x_2$

is $\mathbf{F}_2 = [1, 1, 0, 1]^T$. Since $\mathbf{F}_2 = \mathbf{GF}_1$, where $\mathbf{G} = \begin{bmatrix} 1 & 1 \\ 0 & 1 \end{bmatrix} \otimes \begin{bmatrix} 1 & 0 \\ 0 & 1 \end{bmatrix} = \begin{bmatrix} 1 & 1 & 0 & 0 \\ 0 & 1 & 0 & 0 \\ 0 & 0 & 1 & 1 \\ 0 & 0 & 0 & 1 \end{bmatrix}$

which is a matrix corresponding to LP-operations, the functions f_1 and f_2 are LP-equivalent.

The same can be concluded by a different way. Since $x \oplus y = \overline{x}y \vee x\overline{y} = x \vee y$, iff $x \cdot y = 0$

$$f_2 = \overline{x}_1 \oplus x_1 x_2 = \overline{x}_1 \vee x_1 x_2 = \overline{x}_1 \vee x_2,$$

it follows that f_1 and f_2 are N-equivalent due to $x_1 \leftrightarrow \overline{x}_1$. Since N-classification involves LP-classification rules, f_1 and f_2 are also LP-equivalent.

1.12 The function f_2 can be derived from f_1 if the variable x_1 is negated $x_1 \leftrightarrow \overline{x}_1$. Negation of a variable is both an NPN and LP-classification rule. In matrix form, it can be expressed by $\mathbf{G} = \begin{bmatrix} 0 & 1 \\ 1 & 0 \end{bmatrix}$. The two remaining variables x_2, x_3 are unchanged. Therefore,

$$\mathbf{G} = \begin{bmatrix} 0 & 1 \\ 1 & 0 \end{bmatrix} \otimes \begin{bmatrix} 1 & 0 \\ 0 & 1 \end{bmatrix} \otimes \begin{bmatrix} 1 & 0 \\ 0 & 1 \end{bmatrix} = \begin{bmatrix} 0 & 0 & 0 & 1 & 0 & 0 & 0 & 0 \\ 0 & 0 & 1 & 0 & 0 & 0 & 0 & 0 \\ 0 & 1 & 0 & 0 & 0 & 0 & 0 & 0 \\ 1 & 0 & 0 & 0 & 0 & 0 & 0 & 0 \\ 0 & 0 & 0 & 0 & 0 & 0 & 0 & 1 \\ 0 & 0 & 0 & 0 & 0 & 0 & 1 & 0 \\ 0 & 0 & 0 & 0 & 0 & 1 & 0 & 0 \\ 0 & 0 & 0 & 0 & 1 & 0 & 0 & 0 \end{bmatrix}.$$

The truth-vectors of functions f_1 and f_2 are $\mathbf{F}_1 = [0, 1, 0, 1, 0, 1, 1, 1]^T$ and $\mathbf{F}_2 = [0, 1, 1, 1, 0, 1, 0, 1]^T$. Indeed $\mathbf{F}_1 = \mathbf{GF}_2$.

1.13 The truth-vectors of functions f_1 and f_2 are $\mathbf{F}_1 = [0, 1, 0, 1, 0, 1, 1, 1]^T$ and $\mathbf{F}_2 = [1, 0, 1, 1, 0, 0, 0, 1]^T$. Since $\mathbf{F}_1 = \mathbf{GF}_2$, the matrix $\mathbf{G}$ is

$$\mathbf{G} = \mathbf{G}_1 \otimes \mathbf{G}_2 \otimes \mathbf{G}_3 = \begin{bmatrix} 1 & 0 \\ 0 & 1 \end{bmatrix} \otimes \begin{bmatrix} 1 & 1 \\ 0 & 1 \end{bmatrix} \otimes \begin{bmatrix} 0 & 1 \\ 1 & 0 \end{bmatrix} = \begin{bmatrix} 0 & 1 & 0 & 1 & 0 & 0 & 0 & 0 \\ 1 & 0 & 1 & 0 & 0 & 0 & 0 & 0 \\ 0 & 0 & 0 & 1 & 0 & 0 & 0 & 0 \\ 0 & 0 & 1 & 0 & 0 & 0 & 0 & 0 \\ 0 & 0 & 0 & 0 & 0 & 1 & 0 & 1 \\ 0 & 0 & 0 & 0 & 1 & 0 & 1 & 0 \\ 0 & 0 & 0 & 0 & 0 & 0 & 0 & 1 \\ 0 & 0 & 0 & 0 & 0 & 0 & 1 & 0 \end{bmatrix}.$$

Each product term $\mathbf{G}_i, i = 1, 2, 3$ in the Kronecker product corresponds to an operation over one of the variables. $\mathbf{G}_1$ shows the identity matrix $x_1 \leftrightarrow x_1$. The matrix $\mathbf{G}_2$ shows $x_2 \leftrightarrow 1$, and $\mathbf{G}_3$ shows the negation of the variable $x_3 \leftrightarrow \bar{x}_3$. It is evident that functions f_1 and f_2 are not NPN-equivalent since $x_2 \leftrightarrow 1$ is not in the NPN-classification rules.

1.14   The Walsh spectrum of the given function is

$$\mathbf{W}_{f_1} = \frac{1}{16}[14, 2, 2, -2, 2, -2, -2, 2, 2, -2, -2, 2, -2, 2, 2, -2]^T.$$

Any invariant operation over the inputs or over the output of the function f_1 will result in a function that belongs to the same class. For example, the replacement of the variable $x_1 \leftrightarrow x_1 \oplus x_2$ results in the permutation of the Walsh spectrum as

$$\mathbf{W}_{f_2} = \frac{1}{16}[14, 2, 2, -2, 2, -2, -2, 2, -2, 2, 2, -2, 2, -2, -2, 2]^T.$$

By the application of the inverse Walsh transform and re-encoding $\{0, 1\}$, we obtain the truth-vector of the equivalent function f_2 as

$$\mathbf{F}_2 = [0, 0, 0, 0, 0, 0, 0, 0, 1, 0, 0, 0, 0, 0, 0, 0]^T.$$

To verify, the replacement $x_1 \rightarrow x_1 \oplus x_2$ in f_1 results in $f_2 = (x_1 \oplus x_2)x_2x_3x_4 = x_1x_2x_3x_4 \oplus x_2x_3x_4 = \bar{x}_1x_2x_3x_4$, which has the truth vector as $\mathbf{F}_2$ above.

1.15   Walsh spectra of functions f_1 and f_2 are

$$\mathbf{W}_{f_1} = \frac{1}{16}[8, 0, 0, 0, 0, 0, 0, 8, 4, 4, 4, -4, 4, -4, -4, -4]^T \quad \text{and}$$
$$\mathbf{W}_{f_2} = \frac{1}{16}[6, -2, 2, 2, 2, 2, -2, 6, 6, 6, 2, -6, 2, -6, -2, -2]^T.$$

Fig. A.1 shows WDDs constructed from $\mathbf{W}_{f_1}$ and $\mathbf{W}_{f_2}$.

It is evident that these WDDs share the same distribution of nodes per levels which can be specified by a vector $\mathbf{D} = [d(i)], i = 1, \ldots, n + 1$, whose i-th element is the number of nodes at the i-th level in the WDD. The first element is always 1, since this is the root node in the WDD. The last element is the number of constant nodes. In this example $[1, 2, 4, 3, 4]$. The values of constant nodes are different up to encoding $(8, 0, 4, -4) \rightarrow (6, -2, 2, -6)$.

1.16   For $f(x_1, x_2, x_3) = \bar{x}_1x_2 \vee \bar{x}_3$, the dual function is

$$
\begin{aligned}
f^d(x_1, x_2, x_3) &= \overline{f(\bar{x}_1, \bar{x}_2, \bar{x}_3)} \\
&= \overline{(x_1\bar{x}_2 \vee x_3)} = (x_1\bar{x}_2)\bar{x}_3 = (\bar{x}_1 \vee x_2)\bar{x}_3.
\end{aligned}
$$

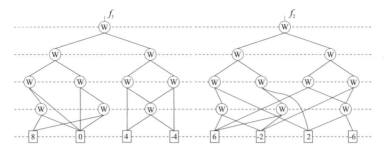

Figure A.1: WDDs for functions f_1 and f_2.

A self-dualized function f^{sd} is defined as

$$f^{sd}(x_1, x_2, x_3, x_4) = x_4 f(x_1, x_2, x_3) \vee \overline{x}_4 f^d(x_1, x_2, x_3)$$
$$= x_4(\overline{x}_1 x_2 \vee \overline{x}_3) \vee \overline{x}_4 \overline{x}_3 (\overline{x}_1 \vee x_2).$$

We recall the statement that a self-dual function of $(n + 1)$ variables and all the possible decompositions into functions of n variables are in the same SD-class. Thus, by assigning the values of any variable to 1 and 0, we get a function in the same SD-class with f and f^{sd}. For instance, if $x_1 = 1$ in f^{sd}, then the equivalent function is $f^{sd}_{x_1=1} = x_4 \overline{x}_3 \vee \overline{x}_4 \overline{x}_3 x_2$.

1.17 In $(1, -1)$ encoding, $\mathbf{F}_1 = [1, 1, 1, 1, 1, 1, 1, 1, 1, 1, 1, -1, 1, 1, 1, 1]^T$, which has the Walsh spectrum $\mathbf{W}_{f_1} = \frac{1}{16}[14, 2, 2, -2, -2, 2, 2, -2, 2, -2, -2, 2, 2, -2, -2, 2]^T$. The linear transformation $x_2 = x_1 \oplus x_2$ permutes the Walsh spectrum
$\mathbf{W}_{f_1} = [r_0, r_4, r_3, r_{34}, r_2, r_{24}, r_{23}, r_{234}, r_1, r_{14}, r_{13}, r_{134}, r_{12}, r_{124}, r_{123}, r_{1234}]^T$ into
$\mathbf{W}_{f_2} = [r_0, r_4, r_3, r_{34}, r_{12}, r_{124}, r_{123}, r_{1234}, r_1, r_{14}, r_{13}, r_{134}, r_2, r_{24}, r_{23}, r_{234}]^T$. The inverse Walsh transform and re-encoding $\{0, 1\}$ produce the truth-vector or the function f_2 as $\mathbf{F}_2 = [1, 1, 1, 1, 1, 1, 1, 1, 1, 1, 1, 1, 1, 1, 1, -1]^T$. To verify, the linear transformation $x_2 = x_1 \oplus x_2$ in $f_1 = x_1 \overline{x}_2 x_3 x_4$ results in

$$f_2 = x_1 \overline{(x_1 \oplus x_2)} x_3 x_4 = x_1 (\overline{x}_1 \oplus x_2) x_3 x_4 = x_1 x_2 x_3 x_4.$$

Notice the difference in the orderings in the Walsh domain produced by the linear transformations $x_1 = x_1 \oplus x_2$ in Exercise 1.14 and $x_2 = x_1 \oplus x_2$. For more detail on this subject, see [22].

A.2   CHAPTER 2

2.1   Consider a PPRM of a bent function. The number of products with degree k is $\binom{n}{k}$. Thus, the number of PPRMs that represent bent functions is

$$\eta(n, k) = \sum_{k=0}^{\frac{n}{2}} \binom{n}{k}.$$

Since

$$\sum_{k=0}^{n} \binom{n}{k} = 2^n,$$

and

$$\binom{n}{k} = \binom{n}{n-k}$$

we have the relation.

2.2

- For $n = 4$.
$$\eta(4) = 2^{4-1} + \frac{1}{2}\binom{4}{2} = 8 + \frac{6}{2} = 11.$$

  Thus, the number of bent functions with $n = 4$ is at most 2^{11}.

- For $n = 6$.
$$\eta(6) = 2^{6-1} + \frac{1}{2}\binom{6}{3} = 32 + \frac{1}{2}\binom{6}{3} = 32 + \frac{20}{2} = 42.$$

  Thus, the number of bent functions with $n = 6$ is at most 2^{42}.

- For $n = 8$.
$$\eta(8) = 2^{8-1} + \frac{1}{2}\binom{8}{4} = 128 + \frac{70}{2} = 163.$$

  Thus, the number of bent functions with $n = 8$ is at most 2^{163}.

2.3   The number of product terms in the PPRM with degree k is $\binom{n}{k}$. To form a homogeneous function, we include or exclude each of the degree-k product terms (2 choices). Excluding all product terms yields the constant 0 function, which does not have degree k. Thus, the number of degree-k homogeneous functions is $2^{\binom{n}{k}} - 1$.

2.4   Sum the number of homogeneous functions of degree-k over $0 \le k \le n$.

2.5

- For $n = 4$.

$$\sum_{k=0}^{4} 2^{\binom{n}{k}} = 2^1 + 2^4 + 2^6 + 2^4 + 2^1 - 4 = 2 + 16 + 64 + 16 + 2 - 4 = 96.$$

- For $n = 6$.

$$\sum_{k=0}^{4} 2^{\binom{n}{k}} = 2^1 + 2^6 + 2^{15} + 2^{20} + 2^{15} + 2^6 + 2^1 - 6 \simeq 2^{20}.$$

- For $n = 8$.

$$\sum_{k=0}^{4} 2^{\binom{n}{k}} = 2^1 + 2^8 + 2^{28} + 2^{56} + 2^{70} + 2^{56} + 2^{28} + 2^8 + 2^1 - 8 \simeq 2^{70}.$$

2.6   In the truth table of a balanced function, there are as many 1's (2^{n-1}) as there are 0's (2^{n-1}). A balanced function is formed by choosing 2^{n-1} entries from 2^n to be 1; the rest are 0's. This can be done in $\binom{2^n}{2^{n-1}}$ ways.

2.7

- For $n = 4$.

$$\binom{2^4}{2^3} = 12870 \simeq 2^{13.65}.$$

- For $n = 6$.

$$\binom{2^6}{2^5} \simeq 2^{60.67}.$$

2.8   If the function used in encrypting a plaintext message is the constant 0 function, then the encoded message is identical to the plaintext message. If the constant 1 function is used instead, then the encoded message is the complement of the plaintext message. In either case, decryption is simple. Note also, that, if the function used in encrypting a plaintext message has nearly all 0's or all 1's, then encryption is somewhat more difficult because there will be "errors", but, with enough plaintext, it is likely that decryption will be ultimately successful. The most difficult keystream to attack is the one with as many 0's as 1's; i.e. the balanced function.

2.9   There are 2^{n+1} affine and 2^{n+1} symmetric functions. The DQF is bent, and there are at least two DQF functions in separate affine classes, for $n \geq 4$. Recall that if f is bent, then so is $f \oplus a$, where a is an affine function. It follows there are at least twice as many bent functions as there are affine functions or symmetric functions.

2.10   The first coefficient s_ϕ in the Walsh spectrum of f is just the sum of 1's and -1's representing 0's and 1's, respectively, in the truth table of f. $s_\phi = w(\bar{f}) = 2^n - 2w(f)$ follows directly by summing $\bar{f} = 1 - 2f$ over all 2^n truth table entries.

2.11   The first coefficient in the Walsh spectrum of f is just the sum of 1's and -1's representing 0's and 1's, respectively in the truth table of f. In a balanced function, the number of 0's is identical to the number of 1's, which means, in the $(1, -1)$ encoding of f, there are an equal number of 1's and -1's. Thus, their sum is 0.

2.12   Outline of the solution to: Prove that a bent function has a flat spectrum. First, by definition, a bent function has maximum nonlinearity $2^{n-1} - 2^{n/2-1}$. Then, assume, on the contrary, that it does not have a flat spectrum. This means that for at least one affine function (row u in W_n), $\widehat{f} \cdot u < 2^{n/2}$ (or $\widehat{f} \cdot u > 2^{n/2}$) (We have to demonstrate that u is a row in W_n expressed in $(1, -1)$ encoding and that it is a linear function. Then, $\widehat{f} \cdot u$ is one spectral coefficient.). Therefore, because $\widehat{f} \cdot u < 2^{n/2}$, this contradicts the assumption that f has maximum nonlinearity $2^{n-1} - 2^{n/2-1}$. Note, that W_n contains only linear functions; you have to accommodate affine functions.

2.13   We use mathematical induction to prove the equation. It is easy to verify that $w(f_4) = 6$. Next, assume that the relation holds for the function with $k = 2r \geq 2$. That is, $w(f_k) = 2^{k-1} - 2^{\frac{k}{2}-1}$, where $k = 2r$. Next, consider the case of $k + 2 = 2(r + 1)$. Since f_{k+2} can be represented as

$$
\begin{aligned}
f_{k+2} &= f_k \oplus x_{k+1}x_{k+2} \\
&= (\bar{x}_{k+1} \vee x_{k+1}\bar{x}_{k+2})f_k \vee x_{k+1}x_{k+2}\bar{f}_m
\end{aligned}
$$

We have

$$
\begin{aligned}
w(f_{k+2}) &= 3w(f_k) + w(\bar{f}_k) = 3w(f_k) + 2^k - w(f_k) \\
&= 2^k + 2w(f_k) = 2^k + 2(2^{k-1} - 2^{\frac{k}{2}-1}) \\
&= 2^{(k+1)} - 2^{\frac{k}{2}} = 2^{(k+2)-1} - 2^{\frac{k+2}{2}-1}
\end{aligned}
$$

Thus, the equation is true for $k + 2$.

2.14   In Problem 12, it was shown that $DQF(x)$ has weight $2^{n-1} - 2^{\frac{n}{2}-1}$. Note that $x_1 \oplus x_1x_2 \oplus x_3x_4 \oplus \ldots \oplus x_{n-1}x_n = x_1\bar{x}_2 \oplus x_3x_4 \oplus \ldots \oplus x_{n-1}x_n$, and from Problem 14, it follows that $x_1 \oplus DQF(x) = DQF(x)|_{\bar{x}_2 \to x_2}$ also has weight $2^{n-1} - 2^{\frac{n}{2}-1}$. Therefore, $DQF(x) \oplus a(x)$ has weight $2^{n-1} \pm 2^{\frac{n}{2}-1}$, where $a(x)$ is an affine function. Thus, $DQF(x)$ is bent.

2.15   The nonlinearity of a function f is the minimum of the distances between f and all affine functions. The distance between f and the affine function 0 is the weight of f. It follows that the nonlinearity of f can be no greater than the weight of f.

2.16   If $f(x)$ is bent, then the minimum distance between $f(x)$ and any affine function is $2^{n-1} - 2^{\frac{n}{2}-1}$. However, $a(x) \oplus a'(x)$, where $a(x)$ and $a'(x)$ are both affine is also affine. It follows that the distance between $f(x) \oplus a(x)$ and the set of affine functions is just a rearrangement of the distances between $f(x)$ and the set of affine functions. It follows that the minimum distance of $f(x) \oplus a(x)$ from the set of affine functions is also $2^{n-1} - 2^{\frac{n}{2}-1}$, and thus, $f(x) \oplus a(x)$ is bent.

2.17   The weight of a function is unchanged when a variable x_i is complemented (simply, the function entries in the truth table are rearranged.). Thus, the weight of $(f(x) \oplus a(x))|_{\bar{x}_i \to x_i} = f(x)|_{\bar{x}_i \to x_i} \oplus a(x)|_{\bar{x}_i \to x_i}$ is the same as the weight of $f(x) \oplus a(x)$. Thus, the distances between $f(x)|_{\bar{x}_i \to x_i}$ and the set of affine functions is the same as the distances between $f(x)$ and the set of affine functions. It follows that the minimum distance between $f(x)|_{\bar{x}_i \to x_i}$ and the set of affine functions is $2^{n-1} - 2^{\frac{n}{2}-1}$. Thus, $f(x)|_{\bar{x}_i \to x_i}$ is bent.

A.3   CHAPTER 3

3.1   Proof of the second inequality of (3.10):

$$f(\mathbf{x}) \leq \max_{x_i} f(\mathbf{x}) \quad \Leftrightarrow \quad f(\mathbf{x}) \wedge \overline{\max_{x_i} f(\mathbf{x})} = 0. \tag{A.3}$$

$$
\begin{aligned}
f(\mathbf{x}) &\wedge \overline{\max_{x_i} f(\mathbf{x})} \\
&= (\bar{x}_i\, f(0) \vee x_i\, f(1)) \wedge \overline{(f(0) \vee f(1))}, \\
&= (\bar{x}_i\, f(0) \vee x_i\, f(1)) \wedge \overline{f(0)} \wedge \overline{f(1)}, \\
&= \bar{x}_i\, f(0)\, \overline{f(0)}\, \overline{f(1)} \vee x_i\, f(1)\, \overline{f(0)}\, \overline{f(1)} = 0.
\end{aligned}
\tag{A.4}
$$

3.2   Proof of (3.7):

$$
\begin{aligned}
\frac{\partial f(\mathbf{x})}{\partial x_i} &= f(x_i, \mathbf{x}_1) \oplus f(\bar{x}_i, \mathbf{x}_1) \\
&= (\bar{x}_i\, f(0) \vee x_i\, f(1)) \oplus (\bar{x}_i\, f(1) \vee x_i\, f(0)), \\
&= (\bar{x}_i\, f(0) \oplus x_i\, f(1)) \oplus (\bar{x}_i\, f(1) \oplus x_i\, f(0)), \\
&= (\bar{x}_i \oplus x_i)\, f(0) \oplus (x_i \oplus \bar{x}_i)\, f(1), \\
&= f(0) \oplus f(1).
\end{aligned}
\tag{A.5}
$$

Proof of (3.8):

$$
\begin{aligned}
\min_{x_i} f(\mathbf{x}) &= f(x_i, \mathbf{x}_1) \wedge f(\bar{x}_i, \mathbf{x}_1), \\
&= (\bar{x}_i\, f(0) \vee x_i\, f(1)) \wedge (\bar{x}_i\, f(1) \vee x_i\, f(0)), \\
&= \bar{x}_i\, f(0)\, f(1) \vee x_i\, f(1)\, f(0), \\
&= (\bar{x}_i \vee x_i)\, (f(0) \wedge f(1)), \\
&= f(0) \wedge f(1).
\end{aligned}
\tag{A.6}
$$

Proof of (3.9):

$$
\begin{aligned}
\max_{x_i} f(\mathbf{x}) &= f(x_i, \mathbf{x}_1) \vee f(\overline{x}_i, \mathbf{x}_1), \\
&= (\overline{x}_i \, f(0) \vee x_i \, f(1)) \vee (\overline{x}_i \, f(1) \vee x_i \, f(0)), \\
&= \overline{x}_i \, f(0) \vee x_i \, f(1) \vee \overline{x}_i \, f(1) \vee x_i \, f(0), \\
&= (\overline{x}_i \vee x_i)(f(0) \vee (\overline{x}_i \vee x_i) f(1)), \\
&= f(0) \vee f(1).
\end{aligned}
\tag{A.7}
$$

3.3 Proof of the first inequality of (3.16):

$$
\min_{\mathbf{x}_0} f(\mathbf{x}_0, \mathbf{x}_1) \leq f(\mathbf{x}_0, \mathbf{x}_1),
$$

$$
\min_{\mathbf{x}_0} f(\mathbf{x}_0, \mathbf{x}_1) \wedge \overline{f(\mathbf{x}_0, \mathbf{x}_1)} = f(\mathbf{x}_0, \mathbf{x}_1) \wedge f(\overline{\mathbf{x}}_0, \mathbf{x}_1) \wedge \overline{f(\mathbf{x}_0, \mathbf{x}_1)} = 0.
\tag{A.8}
$$

Proof of the second inequality of (3.16):

$$
f(\mathbf{x}_0, \mathbf{x}_1) \leq \max_{\mathbf{x}_0} f(\mathbf{x}_0, \mathbf{x}_1),
$$

$$
\begin{aligned}
f(\mathbf{x}_0, \mathbf{x}_1) \wedge \overline{\max_{\mathbf{x}_0} f(\mathbf{x}_0, \mathbf{x}_1)} &= f(\mathbf{x}_0, \mathbf{x}_1) \wedge \overline{f(\mathbf{x}_0, \mathbf{x}_1) \vee f(\overline{\mathbf{x}}_0, \mathbf{x}_1)} \\
&= f(\mathbf{x}_0, \mathbf{x}_1) \wedge \overline{f(\mathbf{x}_0, \mathbf{x}_1)} \wedge \overline{f(\overline{\mathbf{x}}_0, \mathbf{x}_1)} = 0.
\end{aligned}
\tag{A.9}
$$

3.4 Proof of the second inequality of (3.21): from

$$
\max_{x_i} f(\mathbf{x}) \leq \max_{(x_{02}, x_i)}{}^2 f(\mathbf{x}) \ldots \leq \max_{\mathbf{x}_0 \setminus x_{0m}}{}^{m-1} f(\mathbf{x}) \leq \max_{\mathbf{x}_0}{}^m f(\mathbf{x}),
\tag{A.10}
$$

we show without loss of generality

$$
\max_{\mathbf{x}_{0k}}{}^k f(x_{0ik}, \mathbf{x}_{0k}, \mathbf{x}_1) \leq \max_{\mathbf{x}_{0k}, x_{0ik}}{}^{k+1} f(x_{0ik}, \mathbf{x}_{0k}, \mathbf{x}_1).
\tag{A.11}
$$

We substitute $\max_{\mathbf{x}_{0k}}{}^k f(x_{0ik}, \mathbf{x}_{0k}, \mathbf{x}_1) = g(x_{0ik}, \mathbf{x}_1)$ and get due to (3.19):

$$
\begin{aligned}
\max_{\mathbf{x}_{0k}}{}^k f(x_{0ik}, \mathbf{x}_{0k}, \mathbf{x}_1) &\leq \max_{\mathbf{x}_{0k}, x_{0ik}}{}^{k+1} f(x_{0ik}, \mathbf{x}_{0k}, \mathbf{x}_1), \\
g(x_{0ik}, \mathbf{x}_1) &\leq \max_{x_{0ik}} \left[\max_{\mathbf{x}_{0k}}{}^k f(x_{0ik}, \mathbf{x}_{0k}, \mathbf{x}_1) \right], \\
g(x_{0ik}, \mathbf{x}_1) &\leq \max_{x_{0ik}} g(x_{0ik}, \mathbf{x}_1).
\end{aligned}
\tag{A.12}
$$

Due to Theorem 3.2, formula (A.12) proves the last inequality of (3.21).

3.5 Prepare a Boolean space with well ordered variables, the required functions f as object 1, h_0 as object 2, h_1 as object 3, h_2 as object 4, and the tuple of variables $< h_1, h_2 >$.

```
space 32 1
avar 1
x1 x2 x3 x4 h1 h2 f.
sbe 1 1
x2#/(/(x1&/(x2&x3))
&/(x4&/(/(x2&x3)))).
sbe 1 2
x2#/(/(x1&h1)&/(x4&h2)).
sbe 1 3
/(x2&x3).
sbe 1 4
/(/(x2&x3)).
vtin 1 5
h1 h2.
```

Prepare a PRP that calculates the structural dynamic hazards based on the given functions and with regard to the variables prepared as object 10.

```
derv 2 5 11
cpl 11 12
maxk 12 5 13
mink 2 5 14
maxk 2 5 15
syd 14 15 16
derv 3 10 17
derv 4 10 18
derv 1 10 19
isc 13 16 20
isc 20 17 20
isc 20 18 20
isc 20 19 20
```

Execute the PRP with regard to (a) $< x2 >$ as object 10, (b) $< x3 >$ as object 10, and (c) $< x2\,x3 >$ as object 10. The wanted structural dynamic hazards are

$$\text{(a)} \quad (x_1, x_2, x_3, x_4) = (1111), (x_1, x_2, x_3, x_4) = (1011), \quad \text{(A.13)}$$
$$\text{(b)} \quad \text{there is no structural dynamic hazards with regard to } x_3, \quad \text{(A.14)}$$
$$\text{(c)} \quad (x_1, x_2, x_3, x_4) = (1111), (x_1, x_2, x_3, x_4) = (1001). \quad \text{(A.15)}$$

The gate G_c for the function $\overline{x_1 \wedge x_4}$ must be added as shown below to remove the detected structural dynamic hazards.

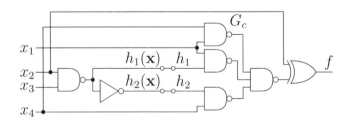

The verification with the PRP given above confirms that the detected structural dynamic hazards are removed.

3.6 From (3.21), we get

$$\min_{\mathbf{x}_b}^l f(\mathbf{x}_a, \mathbf{x}_b, \mathbf{x}_c) \leq f(\mathbf{x}_a, \mathbf{x}_b, \mathbf{x}_c), \tag{A.16}$$

$$\min_{\mathbf{x}_a}^k f(\mathbf{x}_a, \mathbf{x}_b, \mathbf{x}_c) \leq f(\mathbf{x}_a, \mathbf{x}_b, \mathbf{x}_c), \tag{A.17}$$

$$\min_{\mathbf{x}_b}^l f(\mathbf{x}_a, \mathbf{x}_b, \mathbf{x}_c) \wedge \min_{\mathbf{x}_a}^k f(\mathbf{x}_a, \mathbf{x}_b, \mathbf{x}_c) \leq f(\mathbf{x}_a, \mathbf{x}_b, \mathbf{x}_c). \tag{A.18}$$

From (3.3), the inequality (A.18) can be transformed into the equation

$$f(\mathbf{x}_a, \mathbf{x}_b, \mathbf{x}_c) = \min_{\mathbf{x}_b}^l f(\mathbf{x}_a, \mathbf{x}_b, \mathbf{x}_c) \wedge \min_{\mathbf{x}_a}^k f(\mathbf{x}_a, \mathbf{x}_b, \mathbf{x}_c), \tag{A.19}$$

if the condition (3.52) holds. The substitution of (3.54) and (3.55) into (A.19) leads directly to the OR-bi-decomposition (3.43) under the condition that (3.52) holds.

3.7 The single variable b can be replaced by the set of variables $\mathbf{x}_b$ in (3.57) such that we get

$$h(\mathbf{x}_b, \mathbf{x}_c) = \max_a (a \wedge f(a, \mathbf{x}_b, \mathbf{x}_c)). \tag{A.20}$$

3.8 By substituting $h(\mathbf{x}_b, \mathbf{x}_c)$ into (3.44) with $\mathbf{x}_a = a$, we get

$$f(a, \mathbf{x}_b, \mathbf{x}_c) = g(a, \mathbf{x}_c) \oplus \max_a (a \wedge f(a, \mathbf{x}_b, \mathbf{x}_c)), \tag{A.21}$$

$$g(a, \mathbf{x}_c) = f(a, \mathbf{x}_b, \mathbf{x}_c) \oplus \max_a (a \wedge f(a, \mathbf{x}_b, \mathbf{x}_c)). \tag{A.22}$$

3.9 The left-hand side of the equation (A.22) must not depend on $\mathbf{x}_b$ which can be expressed by

$$\Delta_{\mathbf{x}_b} g(a, \mathbf{x}_c) = 0, \tag{A.23}$$

$$\Delta_{\mathbf{x}_b} (f(a, \mathbf{x}_b, \mathbf{x}_c) \oplus \max_a (a \wedge f(a, \mathbf{x}_b, \mathbf{x}_c))) = 0. \tag{A.24}$$

Using (3.1) and definition (3.6), condition (A.24) can be transformed into (3.70) of the EXOR-bi-decomposition with regard to the single variable a and the set of variables $\mathbf{x}_b$:

$$\Delta_{\mathbf{x}_b} (\bar{a} f(a = 0, b, \mathbf{x}_c) \oplus a f(a = 1, b, \mathbf{x}_c) \oplus f(a = 1, b, \mathbf{x}_c)) = 0, \tag{A.25}$$

$$\Delta_{\mathbf{x}_b} (\bar{a} f(a = 0, b, \mathbf{x}_c) \oplus \bar{a} f(a = 1, b, \mathbf{x}_c)) = 0, \tag{A.26}$$

$$\bar{a} \Delta_{\mathbf{x}_b} (f(a = 0, b, \mathbf{x}_c) \oplus f(a = 1, b, \mathbf{x}_c)) = 0, \tag{A.27}$$

$$\Delta_{\mathbf{x}_b} \left(\frac{\partial f(a, b, \mathbf{x}_c)}{\partial a} \right) = 0. \tag{A.28}$$

3.10 The empty TVLs number 13, 14, 20, 21 and 23 as the result of the following PRP indicate that EXOR-bi-decompositions with regard to the pairs (x_1, x_5), (x_1, x_6), (x_3, x_5), (x_3, x_6), and (x_4, x_6) exist.

```
space 32 1
sbe 1 1
(((x1#x2)&x3)+x4)#
((x2&x5)+(x4#x6)).
_derk 1 <x1 x2> 10
_derk 1 <x1 x3> 11
_derk 1 <x1 x4> 12
_derk 1 <x1 x5> 13
_derk 1 <x1 x6> 14
```

```
_derk 1 <x2 x3> 15
_derk 1 <x2 x4> 16
_derk 1 <x2 x5> 17
_derk 1 <x2 x6> 18
_derk 1 <x3 x4> 19
_derk 1 <x3 x5> 20
_derk 1 <x3 x6> 21
_derk 1 <x4 x5> 22
_derk 1 <x4 x6> 23
_derk 1 <x5 x6> 24
```

Using a second PRP, we check whether an EXOR-bi-decomposition with regard to $\mathbf{x}_a = (x_1, x_3)$ and $\mathbf{x}_b = (x_5, x_6)$ exists.

```
tin 1 30
x5 x6.
00.
isc 1 30 31
maxk 31 30 32
```

```
syd 1 32 33
_mink 33 <x1 x3> 34
_maxk 33 <x1 x3> 35
syd 34 35 36
syd 32 35 37
syd 37 1 38
```

$$g(x_1, x_3, x_2, x_4) = (x_1 \oplus x_2) \wedge x_3\, \overline{x}_4 \qquad h(x_5, x_6, x_2, x_4) = \overline{x}_4 \wedge x_2\, x_5 \vee x_6 \wedge \overline{x_2\, x_5}$$

«	»	O	T	TVL 32 (ODA)		
0	0	0	0	0	0
0	1	0	0	0	0
1	1	0	0	0	0
1	0	0	1	0	1
x3	x4	0	1	1	0	x2
		0	0	1	1	x1

«	»	O	T	TVL 35 (ODA)		
0	0	0	0	0	0
0	1	1	1	1	1
1	1	1	1	0	1
1	0	0	0	0	1
x5	x6	0	1	1	0	x4
		0	0	1	1	x2

There is an EXOR-bi-decomposition with regard to $\mathbf{x}_a = (x_1, x_3)$ and $\mathbf{x}_b = (x_5, x_6)$ using the function g (32) and the function h (35) verified by the empty TVL 36 and additionally checked by the empty TVL 38.

3.11 Circuit structure calculated by the several types of bi-decompositions is shown below.

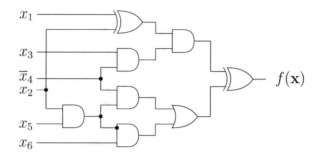

A.4   CHAPTER 4

4.1   A Boolean function over n variables has 2^n input patterns that must be uniquely mapped to the outputs. For this mapping any input-output permutation is allowed. Thus, in total there are $2^n!$ possible reversible functions over n variables.

4.2   The circuit realizes the function shown in Table A.1 and has quantum cost of 14 (from the first gate to the last gate: 1, 5, 1, 1, 1, and 5).

Table A.1: Solution for Exercise 4.2.

a	b	c	a'	b'	c'
0	0	0	1	1	1
0	0	1	0	0	0
0	1	0	0	0	1
0	1	1	0	1	1
1	0	0	1	0	0
1	0	1	0	1	0
1	1	0	1	1	0
1	1	1	1	0	1

4.3   The *AND* function is irreversible, since

- its number of inputs differs from its number of outputs and

- there is no unique input-output mapping (i.e., the output pattern 0 occurs three times).

A possible embedding in a reversible function is shown in Table A.2.

Table A.2: Solution for Exercise 4.3.					
0	a	b	o	–	–
0	0	0	0	0	0
0	0	1	0	0	1
0	1	0	0	1	0
0	1	1	1	0	0
1	0	0	0	1	1
1	0	1	1	0	1
1	1	0	1	1	0
1	1	1	1	1	1

4.4 Using the MMD approach to synthesize the function from Table 4.9, the circuit depicted in Figure A.2 results.

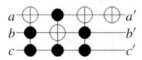

Figure A.2: Solution for Exercise 4.4.

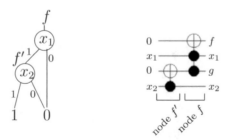

Figure A.3: Solution for Exercise 4.5.

4.5 Figure A.3(a) shows the BDD representing the *AND* function. The circuit resulting from that BDD is given in Figure A.3(b).

A.5   CHAPTER 5

5.1 $||D|| = 26$.

5.2 $tail(\{2, 4, 5\}) = 5$.

5.3 T_1, T_3, T_4, T_6, T_8, T_9, and T_{11}.

5.4 $\{ab, bc, a, b, c, \lambda\}$. Same as for $\theta = 7$.

5.5

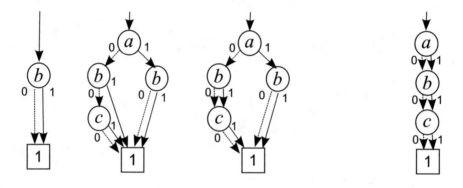

5.6

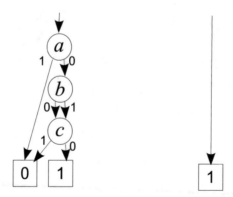

For the set of sparse itemsets, ZDDs are more efficient than BDDs. BDD is better for the set of all the itemsets.

5.7   We just show the trace of subroutine calls.

call LCMovZDD(λ)
    call LCMovZDD(c)
        call LCMovZDD(bc)
        return $\{\lambda\}$
    return $\{\lambda, b\}$
    call LCMovZDD(b)
        call LCMovZDD(ab)
        return $\{\lambda\}$
    return $\{\lambda, a\}$
    call LCMovZDD(a)
    return $\{\lambda\}$
return $\{\lambda, c, bc, b, ab, a\}$

5.8

Table A.3:	
Database name	**Ave. $|T_i|/|\mathcal{E}|$ (%)**
mushroom	20.18
BMS-WebView-1	0.051
BMS-WebView-2	0.138
T10I4D100K	1.161
chess	49.33
connect	33.33
pumsb	3.502

$||D||$ divided by #Transactions gives the average of $|T_i|$. Then, we divide it by #Items. As shown in the above results, we can see examples of practical databases that consist of the sparse itemsets. In such cases, ZDDs are much more effective than ordinary BDDs.

5.9 Frequent itemsets $F = \{ab, bc, a, b, c, \lambda\}$. We compute $S \leftarrow F.\text{onset}(b).\text{change}(b)$, then $S = \{ab, bc, b\}$. Next, $S \leftarrow F.\text{offset}(c)$ gives the result: $S = \{ab, b\}$.

5.10

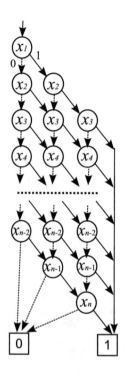

5.11   Generate two ZDDs, F_1 for minimum support $\theta = 7$ and F_2 for minimum support $\theta = 10$. Then, compute a ZDD for $F_1 \setminus F_2$.

Index

Authors' Biographies

RADOMIR S. STANKOVIĆ

Radomir S. Stanković received B.Sc. degree in electronic engineering from the Faculty of Electronics, University of Niš, Serbia, in 1976 and M.Sc. and Ph.D. degrees in applied mathematics from the Faculty of Electrical Engineering, University of Belgrade, Belgrade, Serbia, in 1984, and 1986, respectively. Currently, he is Professor at the Department of Computer Science, Faculty of Electronics, University of Niš, Niš, Serbia. In October to December 1997, he was a visiting researcher at the Sasao Lab, Department of Computer Science and Electronics, Kyushu Institute of Technology, Iizuka, Japan. From 1999, he is working in part at the Department of Signal Processing, Tampere University of Technology, Tampere, Finland, where he is currently an adjunct professor. His research interests include spectral techniques, switching theory, multiple-valued logic, and signal processing.

STANISLAV STANKOVIĆ

Stanislav Stanković received his M.Sc. from Faculty of Electronics, University of Niš, Niš, Serbia in 2003 and Ph.D. degree from Faculty of Computing and Electrical Engineering, Tampere University of Technology, Tampere, Finland, in 2009. From 2003 to 2004 he was with the Computer Vision Laboratory at École Politechnique Fédérale de Lausanne, Lausanne, Switzerland. Since 2005, he is a researcher at the Department of Signal Processing, Tampere University of Technology, Tampere, Finland. His research interests are programming, data representations, signal processing, and switching theory.

HELENA ASTOLA

Helena Astola is an M.Sc. student in Information Technology at Tampere University of Technology, Tampere, Finland. She is a research assistant at the Department of Signal Processing, Tampere University of Technology.

JAAKKO ASTOLA

Jaakko Astola received his Ph.D. degree in mathematics from Turku University, Finland, in 1978. From 1976 to 1977, he was with the Research Institute for Mathematical Sciences of Kyoto University, Kyoto, Japan. Between 1979 and 1987, he was with the Department of Information Technology, Lappeenranta University of Technology, Lappeenranta, Finland. In 1984, he worked as a visiting scientist in Eindhoven University of Technology, The Netherlands. From 1987 to 1992, he was Associate Professor in Applied Mathematics at Tampere University, Tampere, Finland. From 1993, he has been Professor of Signal Processing at Tampere University of Technology and is currently head of the Academy of Finland Centre of Excellence in Signal Processing leading a group of about 80 scientists. His research interests include signal processing, coding theory, switching theory, spectral techniques, and statistics. He is a Fellow of the IEEE.

JON T. BUTLER

Jon T. Butler received the B.E.E. and M.Engr. degrees from Rensselaer Polytechnic Institute, Troy, New York, U.S.A., in 1966 and 1967, respectively. He received the Ph.D. degree from The Ohio State University, Columbus, Ohio, U.S.A., in 1973. Since 1987, he has been a Professor at the Naval Postgraduate School, Monterey, California, U.S.A. From 1974 to 1987, he was at Northwestern University, Evanston, Illinois, U.S.A. During that time he served two periods of leave at the Naval Postgraduate School, first as a National Research Council Senior Postdoctoral Associate (1980-1981) and second as the NAVALEX Chair Professor (1986-1987). He served one period of leave as a foreign visiting professor at the Kyushu Institute of Technology, Iizuka, Japan. His research interests include logic optimization, multiple-valued logic, and reconfigurable computing. He has served on the editorial boards of the *IEEE Transactions on Computers, Computer*, and the *IEEE Computer Society Press*. He has served as the editor-in-chief of *Computer* and the *IEEE Computer Society Press*. He received the Award of Excellence, the Outstanding Contributed Paper Award, and a Distinctive Contributed Paper Award for papers presented at the International Symposium on Multiple-Valued Logic. He received the Distinguished Service Award, two Meritorious Awards, and nine Certificates of Appreciation for service to the IEEE Computer Society. He is a Life Fellow of the IEEE.

TSUTOMU SASAO

Tsutomu Sasao received the B.E., M.E., and Ph.D. degrees in Electronics Engineering from Osaka University, Osaka Japan, in 1972, 1974, and 1977, respectively.

He has held faculty/research positions at Osaka University, Japan, IBM T. J. Watson Research Center, Yorktown Height, NY and the Naval Postgraduate School, Monterey, CA. He has served as the Director of the Center for Microelectronic Systems at the Kyushu Institute of Technology, Iizuka, Japan. Now, he is a Professor of Department of Computer Science and Electronics. His research areas include logic design and switching theory, representations of logic functions, and multiple-valued logic. He has published more than 8 books on logic design including, Logic Synthesis and Optimization, Representation of Discrete Functions, Switching Theory for Logic Synthesis, and Logic Synthesis and Verification, Kluwer Academic Publishers 1993, 1996, 1999, 2002 respectively. He has served as Program Chairman for the IEEE International Symposium on Multiple-Valued Logic (ISMVL) many times. Also, he was the Symposium Chairman of the 28th ISMVL held in Fukuoka, Japan in 1998. He received the NIWA Memorial Award in 1979, Takeda Techno-Entrepreneurship Award in 2001, and Distinctive Contribution Awards from IEEE Computer Society MVL-TC for papers presented at ISMVLs, in 1987, 1996, 2003. He has served an associate editor of the *IEEE Transactions on Computers*. He is a Fellow of the IEEE.

BERND STEINBACH

Bernd Steinbach studied Information Technology at the University of Technology in Chemnitz (Germany) and graduated with an M.Sc. in 1973. He graduated with a Ph.D. and with a Dr. sc. techn. (Doctor scientiae technicarum) for his second doctoral thesis from the Faculty of Electrical Engineering of the Chemnitz University of Technology in 1981 and 1984, respectively. In 1991 he obtained the Habilitation (Dr.-Ing. habil.) from the same Faculty.

He was working in industry as an Electrician, where he tested professional controlling systems at the Niles Company. After his studies, he taught as an Assistant Lecturer at the Department of Information Technology of the Chemnitz University of Technology. In a following period of industrial occupation as research engineer, he developed programs for test pattern generation for computer circuits at the company Robotron. He returned to the Department of Information Technology of the Chemnitz University of Technology as an Associate Professor for design automation in logic design. Since 1992, he is a Full Professor of Computer Science / Software Engineering and Programming at the Freiberg University of Mining and Technology, Department of Computer Science. He has served as head of the Department of Computer Science and Vice-Dean of the Faculty of Mathematics and Computer Science.

His research areas include logic functions and equations and their application in many fields, such as Artificial Intelligence, UML-based testing of software, UML-based hardware/software co-design. He is the head of a group that developed the XBOOLE software system. He published three books in logic synthesis. The first one, co-authored by D. Bochmann, covers Logic Design using XBOOLE (in German). The following two, co-authored by Christian Posthoff, are "Logic Functions and Equations – Binary Models for Computer Science" and "Logic Functions and Equations – Examples and Exercises", Springer 2004, and 2009, respectively. He published more than 170 chapters in books, complete issues of journals, and papers in journals and proceedings.

He has served as Program Chairman for the IEEE International Symposium on Multiple-Valued Logic (ISMVL), and as guest editor of the Journal of Multiple-Valued Logic and Soft Computing. He is the initiator and general chair of the two annual series of the International Workshops on Boolean Problems (IWSBP), which started in 1994 with 8 workshops until now.

He received the Barkhausen Award from the University of Technology Dresden in 1983.

CHRISTIAN POSTHOFF

Christian Posthoff studied Mathematics at the University of Leipzig (Germany) and graduated with an M.Sc. in 1968. He graduated with a Ph.D. in Mathematics in 1975 and with a Dr. sc.

techn. (Doctor scientiae technicarum) for his second doctoral thesis from the Faculty of Electrical Engineering of the Chemnitz University of Technology (Germany) in 1979.

He was working in industry as a programmer from 1968 until 1972 and started in 1972 as an Assistant Lecturer at the Department of Information Technology of the Chemnitz University of Technology. In 1972, he was promoted to the position of Associate Professor for logic design. In 1983, he started as Full Professor of Computer Science at the Department of Computer Science, which has been set up at this point of time. In 1994, he became the Chair of Computer Science at the University of The West Indies, St. Augustine (Trinidad & Tobago). From 1996 to 2002, he was the Head of the Department of Mathematics & Computer Science.

His main research activities concentrated on the investigation of the applicability of classic and non-classic logics in Computer Science, from the theoretical base to algorithmic and programmable concepts. An independent direction of research activities within AI, investigations of computer chess and other strategic games, arose from his love of chess. His actual research activities mainly concern learning from examples, the construction of intelligent tutoring systems, the parallelization of inference mechanisms, systems of diagnosis and configuration. In cooperation with colleagues from mechanical engineering and medicine, he has been supervising the construction of several expert systems.

He has served as a referee for many conferences and journals and is the author of 15 books and more than 100 publications in different journals, conference volumes etc. His two last books are co-authored with B. Steinbach, "Logic Functions and Equations - Binary Models for Computer Science" and "Logic Functions and Equations - Examples and Exercises", Springer 2004, and 2009, respectively.

Four times he received the Scientific Award of the Chemnitz University of Technology. In 2001, he received the Vice-Chancellor's Award of Excellence at the University of the West Indies.

ROBERT WILLE

Robert Wille received the Diploma and Dr.-Ing. degrees in computer science from the University of Bremen, Bremen, Germany, in 2006 and 2009, respectively. He is currently with the Group of Computer Architecture at the University of Bremen. His research interests include reversible logic and quantum computation, techniques for solving satisfiability problems, as well as hardware verification. Dr. Wille was a recipient of the 2008 Young Researchers Award from the International Symposium on Multiple-Valued Logic.

ROLF DRECHSLER

Rolf Drechsler received the Diploma and Dr. Phil. Nat. degrees in computer science from J. W. Goethe University, Frankfurt am Main, Germany, in 1992 and 1995, respectively. From 1995 to 2000, he was with the Institute of Computer Science, Albert Ludwigs University, Freiburg, Germany. In 2000, he was a Senior Engineer with the Formal Verification Group, Corporate Technology, Siemens

AG, Munich, Germany. Since October 2001, he has been with the University of Bremen, Bremen, Germany, where he is currently a Full Professor of computer architecture. He has authored more than ten books in the area of VLSI CAD and published more than 250 papers in archival journals and refereed conference proceedings. His research interests include verification, testing, and synthesis.

Dr. Drechsler was the recipient of Best Paper Awards at the 2006 Haifa Verification Conference and the 2007 Forum on Specification and Design Languages. He was a member of the Program Committee of major conferences in Electronic Design Automation, including the Conference on Computer-Aided Design, the Design Automation Conference, the Asia and South Pacific Design Automation Conference, and the Design Automation and Test Europe.

SHIN-ICHI MINATO

Shin-ichi Minato is an Associate Professor of Graduate School of Information Science and Technology, Hokkaido University, Hokkaido, Japan. He also serves a Project Director of ERATO (Exploratory Research for Advanced Technology) MINATO Discrete Structure Manipulation System Project, executed by JST (Japan Science and Technology Agency). He received the B.E., M.E., and D.E. degrees from Kyoto University, Kyoto, Japan in 1988, 1990, and 1995, respectively. He had been working at NTT Laboratories since 1990 until March 2004. He was a Visiting Scholar at Computer Science Department of Stanford University in 1997. He was a Senior Researcher of NTT Network Innovation Laboratories in 1999. He joined Hokkaido University in 2004. He started the ERATO Project from Oct. 2009.

His research topics include efficient representations and manipulation algorithms for large-scale discrete structure data. He published "Binary Decision Diagrams and Applications for VLSI CAD" (Kluwer,1995). He proposed a data structure "ZDD" (Zero-suppressed BDD) in 1993, which is included in Knuth's book "The Art of Computer Programming" (Vol. 4, Fascicle 1, 2009). He is a member of IEEE, IEICE, IPSJ, and JSAI.

Printed in the United States
by Baker & Taylor Publisher Services